ENVIRONMENTAL LEADERS AND LAGGARDS IN EUROPE

ASHGATE STUDIES IN ENVIRONMENTAL POLICY AND PRACTICE

Series Editor: Adrian McDonald

Based on the Avebury Studies in Green Research series, this wide-ranging series still covers all aspects of research into environmental change and development. It will now focus primarily on environmental policy, management and implications (such as effects on agriculture, lifestyle, health etc). It includes both innovative theoretical research and international practical case studies.

Environmental Leaders and Laggards in Europe

Why there is (not) a 'Southern Problem'

TANJA A. BÖRZEL
Humboldt University Berlin

Routledge
Taylor & Francis Group

LONDON AND NEW YORK

First published 2003 by Ashgate Publishing

Published 2017 by Routledge
2 Park Square, Milton Park, Abingdon, Oxfordshire OX14 4RN
711 Third Avenue, New York, NY 10017, USA

First issued in paperback 2017

Routledge is an imprint of the Taylor & Francis Group, an informa business

British Library Cataloguing in Publication Data
Boerzel, Tanja
 Environmental leaders and laggards in Europe : why there is
 (not) a 'southern problem'. - (Ashgate studies in
 environmental policy and practice)
 1. Environmental policy - European Union countries -
 Cross-cultural studies 2. Compliance 3. Environmental law -
 European Union countries
 I. Title
 333. 7 ' 094

Library of Congress Cataloging-in-Publication Data
Börzel, Tanja A.
 Environmental leaders and laggards in Europe : why there is (not) a 'southern problem' /
Tanja Boerzel.
  p. cm. -- (Ashgate studies in environmental policy and practice)
 Includes bibliographical references and index.
 ISBN 0-7546-1687-8
 1. Environmental policy--Europe. I. Title. II. Series.

 GE212 .B67 2003
 363.7'056'094--dc21   2002028164

ISBN 13: 978-1-138-25816-7 (pbk)
ISBN 13: 978-0-7546-1687-0 (hbk)

Contents

List of Figures, Tables and Charts

Preface and Acknowledgements

This book would never have been written if Adrienne Héritier, my PhD advisor, had not insisted that I should give my claims about the transformation of the state some 'real' empirical grounding. In her irresistible way, she talked me into doing a policy study, and I chose the environment, which appeared to me the least boring. A few weeks later, I found myself in Barcelona visiting local water plants in order to understand the problems of Spanish regions in implementing the European Drinking Water Directive. I will never forget my first international conference where I presented the results of my policy study – people's eyes glazed over when I told them about how difficult it was for Catalan municipalities to comply with European water policies.

I am most grateful to Alberta Sbragia, who read through the first results of my case studies and strongly encouraged me to go on with my work. Thanks to Christoph Knill and Andrea Lenschow, who took me into their project on the implementation of EU environmental policies, I learned how to link detailed empirics with broader theoretical issues. The project helped me to relate my empirical case studies back to the overall question of my dissertation on how Europeanization affects the domestic structures of the member states.

After two years of laborious field research, I naturally included all the empirical details in my PhD thesis. I had a hard job convincing the Political Science Department of the European University Institute to accept the final draft, which was 50,000 words over the limit. Of course, no press was willing to publish a manuscript with more than 400 pages and some 900 footnotes. Cambridge made me cut one-third of my thesis before they would even send it out to the reviewers. It was the policy study, in which they were the least interested. At this point, I decided to do a separate book, which would allow me to fully exploit the assets of my 12 case studies. I felt that my German-Spanish comparison could make a valuable contribution to an ongoing debate in the field of EU environmental policy-making on whether the EU is having a 'Southern' problem and whether its four southern member states are suffering from a disease called 'Mediterranean Syndrome'. After having extensively studied the German case, it became clear to me that non-compliance with European Environmental Law could not be simply blamed on the southern member states. What was needed was an approach that could explain implementation failure and non-compliance across the alleged North-South divide.

The empirical study draws on various sources. Most of the data were collected as part of the already mentioned research project by Christoph Knill and Andrea Lenschow. I am thankful to them for coordinating the project and to the European Commission, which funded it. The statistical data were drawn from a database, which I constructed during my time as the Coordinator of Environmental Studies at the Robert Schuman Centre for Advanced Studies of the European University

Institute. I am particularly indebted to Yves Mény, the former director of the Schuman Centre, for providing the funding for my research assistant, Charalampos (Babis) Koutalakis, who entered more than 20,000 infringement cases into the database. Without Babis' dedication and endurance, the database would not have come into existence. While we ultimately abandoned it, Babis' work was not entirely wasted – the Commission would have never given us access to its database, if we had not constructed our own in the first place.

Susan Baker, Matthijs Boogarts, Sonja Bugdahn, Jeffrey Checkel, Alf-Inge Jansen Manuel Jiminez, Christian Joerges, Andrew Jordan, Christoph Knill, Maria Kousis, Andrea Lenschow, Andrea Liese, Duncan Liefferink, Matthias Maier, Yves Mény, Fritz Scharpf, Cornelia Ulbert, Dieter Wolf, and Michael Zürn, commented on various parts of the book. My thanks go to all of them. The presentations in the Working Group of Environmental Studies at the Robert-Schuman-Centre for Advanced Studies and the workshop on 'Coming to Terms with the Mediterranean Syndrome', organized at the European University Institute in May 2000, were particularly important for me to test my theoretical argument and discuss the empirical findings of my study. I am particularly grateful to the participants of the 'compliance workshop' organized by Christian Joerges and Michael Zürn at the ECPR Joint Session of Workshops in Mannheim, in March 1999.

Finally, Thomas read through the whole manuscript twice. As an International Relations scholar he has never developed a great taste for my detailed empirics. But his reservations encouraged me to look for 'the big picture' in my stories and link them to broader theoretical questions.

The book is dedicated to all students who seek to combine theory-guided research with thorough empirical case studies.

Tanja A. Börzel

List of Abbreviations

ADENA	Asociación para la Defensa de la Naturaleza (WWF)
AEDENAT	Asociación Ecologista para la Defensa de la Naturaleza
AENOR	Asociación Española de Normalización
AI	Access to Information
Art.	Article
BAT(NEEC)	Best Available Technology (Not Entailing Excessive Costs)
BBU	Bundesverband Bürgerinitiativen Umweltschutz
BGBl.	Bundesgesetzblatt
BImSchG	Bundesimmissionsschutzgesetz
BMU	Bundesministerium für Umwelt, Naturschutz und Reaktorsicherheit
BOE	Boletín Oficial de España
BR-Drs.	Bundesrat-Drucksache
BT-Drs.	Bundestag-Drucksache
BUND	Bund für Umwelt und Naturschutz Deutschland
CAMA	Consejo Asesor de Medio Ambiente
CE	Constitucón Española
CEOE	Confederación Española de Organizaciones Empresariales
CODA	Confederación Española de la Defensa Ambiental
DAZU	Deutsche Akkreditierungs- und Zulassungsgesellschaft für Umweltgutachter mbh
DEPANA	Liga per la Defensa del Patrimoni Natural
DIN	Deutsches Institut für Normung
DNR	Deutscher Naturschutz Ring
DOGC	Diario Oficial de la Generalitat de Catalunya
EC	European Community
ECJ	European Court of Justice
EEC	European Economic Community
EIA	Environmental Impact Assessment
EIS	Environmental Impact Studies
EMAS	Eco-Management and Audit Scheme
ENAC	Entidad Nacional de Acreditación
EU	European Union
EuGH	Europäischer Gerichtshof
EUR	Euro
GDP	Gross Domestic Product
GFAV	Großfeuerungsanlagen-Verordnung
GG	Grundgesetz
IPPC	Integrated Pollution Prevention and Control

ISPA	Instrument for Structural Policies for Pre-Accession
ISO	International Standardization Organization
LCP	Large Combustion Plant
MIMA	Ministerio de Medio Ambiente
NABU	Deutscher Naturschutzbund
NGO	Non-governmental Organization
NIMBY	'not in my backyard'
NPI	New Policy Instruments
OECD	Organization of Economic Cooperation and Development
OJ	Official Journal of the European Communities
PHARE	Poland and Hungary Action for Restructuring of the Economy
RAMINP	Reglamiento de Actividades Molestas, Insalubres, Nocivas y Peligrosas
RD	Real Decreto
SRU	Rat der Sachverständigen für Umweltfragen
STC	Sentencia de Tribunal Constitucional
TA	Technische Anleitung
UAG	Umweltauditgesetz
UBA	Umweltbundesamt
UGA	Umweltgutachterausschuss
UIG	Umweltinformationsgesetz
UMK	Umweltministerkonferenz
UNE	Una Norma Española
UVPG	Umweltverträglichkeitsprüfungsgesetz
VwVfG	Verwaltungsverfahrensgesetz
WHG	Wasserhaushaltsgesetz
WHO	World Health Organization
WWF	World Wildlife Fund

Chapter One

Introduction

No other European laws are so frequently violated as environmental Directives. Over 20 per cent of the infringements registered with the European Commission fall into the area of environment. While compliance is rather poor in general, the four southern member states have the reputation of being particular laggards. Italy, Greece, Portugal, and Spain account for about 42 per cent of the infringement proceedings initiated since 1986. The literature therefore tends to treat ineffective implementation and non-compliance with European environmental policy as a peculiar 'Southern problem' (Pridham and Cini 1994; Pridham 1996). The poor compliance record of the Southern European countries is attributed to certain defects of their political and social institutions. Insufficient administrative capacity, a civic culture inclined to individualism, clientelism, and corruption, and a fragmented, reactive and party-dominated policy process are believed to undermine their willingness and ability to comply with EU environmental law.

This book challenges the view that the Southern European member states suffer from a common disease some have referred to as the 'Mediterranean Syndrome' (La Spina and Sciortino 1993), which makes them the main culprits of non-compliance with EU environmental law. Such a view reproduces specific Northern European images of Southern European politics and ignores general causes of implementation failure and non-compliance. It overlooks significant variations both among the Southern European member states themselves as well as between them and their Northern European counterparts. Not only do Italy, Spain, Greece, and Portugal substantially differ in their political and social institutions as well as in their compliance performance. Northern European countries may face serious compliance problems, too. It took Germany almost ten years and a conviction by the European Court of Justice to correctly transpose the Drinking Water Directive into national law. Spain, on the contrary, had legally implemented the Directive four years before it joined the European Community. Germany has been a front-runner in air pollution control, while Spain is still lagging behind. But Germany has faced as many infringement proceedings as Spain for resisting the implementation of 'new' policy instruments, such as the Environmental Impact Assessment Directive and the Access to Information Directive.

How can we explain variations in member-state compliance with EU environmental law, which cut across the alleged North-South divide? Why do member states successfully implement some policies while leaving others insufficiently transposed, applied and enforced? Are some policies more conducive to effective implementation and compliance than others? Do certain institutional factors facilitate or prohibit the successful integration of European policies into national regulatory structures?

The book attempts to establish two major claims in addressing these questions. First, effective implementation and compliance with EU environmental policies vary across both member states and policies. Therefore, neither the domestic institutions of the member states nor specific features of EU environmental policy-making can account for compliance problems only. Non-compliance results from an interplay between European and domestic factors. If an EU policy does not fit the regulatory structures in a member state, its legal transposition, practical application, and enforcement impose considerable costs of adaptation, which domestic actors are hardly inclined to bear. Laws and administrative procedures have to be changed, and with them, policy-makers and administrators have often to adapt their beliefs in what is the best way of protecting the environment. The integrated approach of the Environmental Impact Assessment Directive, for instance, not only requires comprehensive legal and administrative changes in member states with traditionally fragmented regulatory structures, like Germany. It also challenges the media-specific approach to environmental regulation, which corresponds to a general belief in the technical superiority of bureaucratic organization through specialization. Moreover, resources have to be (re)allocated in order to provide the additional personnel, expertise, and technology for the effective application, monitoring, and enforcement of the new policy. The monitoring of the 64 parameters prescribed by the European Drinking Water Directive required new measurement technologies even for countries like Germany, whose monitoring system was already up to the highest standards. Regulated parties, finally, are likely to oppose 'misfitting' policies, when they impose new regulatory burdens on them. German industry has mobilized against the Access to Information Directive, since it feared that a greater involvement of the public would slow-down the authorization of classified activities. In view of such costs, European policies that do not fit policies at the domestic level are most likely to result in implementation failure and non-compliance. If, by contrast, a European policy is compatible with the dominant problem-solving approach, the policy instruments and policy standards of a member state, there is no reason why implementation and compliance should give rise to substantial problems.

Second, policy misfit is only the necessary cause of implementation failure and non-compliance. While policy-makers, administrators and regulated parties are likely to resist the costs of complying with 'misfitting' policies, additional pressure from 'below' and from 'above' (Brysk 1993) can induce policy-makers and public authorities to effectively implement and enforce costly European policies. Pressure from below is generated by the mobilization of domestic actors seeking to pull down European policies to the domestic level. Citizens, environmental organizations, and political parties can mobilize public opinion by raising concerns about the non-compliance with a European policy. Media campaigns and public shaming are particularly effective in countries like Germany, which aspire to be environmental leaders in Europe. For the red-green coalition government of Chancellor Schröder, it was increasingly difficult to justify why – after ten years – Germany still did not comply with 'progressive' environmental policies such as the Access to Information Directive or the Environmental Impact Assessment Directive. Moreover, societal actors can act as 'watchdogs' collecting and distributing evidence on instances of non-compliance. Spanish and German environmental organizations have systematically tested the information rights granted by the Access to Information Directive and published the results in several studies. The supremacy and direct effect of European Law also allows individuals to bring legal action against compliance failures. Citizens and public interest groups have successfully litigated against the authorization of projects in German and Spanish courts that had proceeded without the necessary environmental impact assessment. But not only societal actors have an incentive to pull European policies down to the domestic level. Economic actors may also mobilize in favour of compliance with 'misfitting' policies, if they anticipate competitive disadvantages. German business associations put significant pressure on the German government to speedily implement the European Eco-Audit Regulation because they feared that their competitors could gain a first-mover advantage. If domestic mobilization generates sufficient pressure from 'below', it can change the cost-benefit calculations of policy-makers and public authorities making non-compliance more costly.

Domestic mobilization is particularly effective if it succeeds in generating pressure from 'above'. The European Commission has the power to open infringement proceedings against member states which it suspects of violating European Law. Infringement proceedings provide the Commission with an effective means to 'push' member states into compliance. Not only can they lead to a conviction by the European Court of Justice, which may impose substantial financial penalties. Infringement proceedings also incur significant political costs on member states, particularly if they wish to portray themselves as environmental leaders or good Europeans. The German public was rather irritated when the European Commission threatened Germany with a daily fine of over EUR 200,000 for its persistent non-compliance with the Environmental Impact Assessment Directive. Domestic actors fulfil important monitoring functions for the Commission since it has very limited resources. Most infringement proceedings originate with complaints lodged by citizens, public interest groups, and companies. The European infringement proceedings also offer domestic actors an authoritative venue to challenge non-compliant state behaviour. While Spanish and German courts have tended to support the restrictive application of the Access to Information Directive by public authorities, the European Court of Justice has backed appeals by German and Spanish citizens and public interest groups. If member states become 'sandwiched' between pressure from below, where domestic actors pull the EU policy down to the domestic level, and from above, where the Commission and the European Court of Justice push towards compliance, European policies are more likely to be effectively implemented and complied with, despite the high costs involved.

This 'Pull-and-Push' Model of compliance generates the following hypothesis about when 'misfitting' policies are likely to be effectively implemented and complied with despite the adaptational costs they incur:

Compliance with policies that incur significant adaptational costs (policy misfit), is the more likely, the higher the adaptational pressure from below and from above (domestic mobilization, infringement proceedings).

Linking policy and country variables in explaining implementation failure and noncompliance allows to reformulate the 'Southern Problem' in more general terms. Non-compliance with European Environmental Law is clearly more pronounced in the South than in the North. But the reason for this is not some cultural and institutional deficiencies. Rather, EU environmental policy-making produces greater policy misfit for Southern European member states while their closed political opportunity structures and lower level of socio-economic development discourage domestic mobilization that is able to generate pressure from below and from above. EU environmental policy-making is a regulatory contest between high-regulating member states (Héritier, Knill, and Mingers 1996). The industrial and environmental first-comers of the North have developed strict environmental regulations, which they wish to harmonize at the European level to avoid competitive disadvantages for their industries and to reduce adaptational costs in the implementation of European policies. They command the policies as well as the necessary resources to upload them to the European level. The industrial and environmental late-comers of the South, by contrast, have neither the policies nor the capacity for up-loading. They wish to maintain their lower level of regulation to catch up with the industrially more advanced countries as much as they seek to avoid the costs for adapting their regulatory structures to more stringent European standards. The dynamics of European decision-making prevent individual member states to systematically impose their regulatory approach on the others. Since the highly-regulating countries differ significantly in their regulatory structures, even environmental first-comers can face significant policy misfit. In general, however, the southern member states are much more likely to have to down-load European environmental policies that do not fit their regulatory structures. As a result, those countries with the most limited compliance capacities have to bear the highest compliance costs. European policies that are shaped by the economic interests and environmental concerns of the highly industrialized North undermine both the capacity and the willingness of southern member states to comply. But not only do the southern countries have to cope with the problem of few resources and high compliance costs. Their policy-makers and administrators are also less likely to face pressure from domestic actors, which may help to counteract attempts of avoiding compliance costs. The closed political opportunity structures combined with a lower level of socio-economic development seriously constrain the capacity of domestic actors in southern societies to pull European policies down to the domestic level. Limited resources, restricted access to the policy process, and a strong political priority for economic development over environmental protection make it difficult for citizens and public interests groups to generate sufficient pressure at the domestic and European level to significantly increase the costs of non-compliance for policy-makers and administrators.

Explaining implementation and compliance as the result of differing degrees of policy misfit and of pressure from below and above has several advantages over alternative explanations. The 'Pull-and-Push' Model systematically links country and policy variables in accounting for the different levels of compliance in the member states. It can explain why some countries face fewer difficulties in implementing and complying with European (Environmental) Law than others. At the same time, the model explains why individual member states comply better with some policies than with others. Most importantly, the 'Pull-and-Push' Model allows to overcome the reductionism of the 'Southern Problem' literature by emphasizing the general causes of implementation failure and non-compliance, which are neither geographically bound nor static. The capacity to shape European policies, on the one hand, and the openness of the political opportunity structures and the level of socio-economic development, on the other, vary between all member states. They may also change over time. While there seems to be no cure for the 'Mediterranean Syndrome', strengthening the administrative capacities of the southern member states and the organizational capacities of their domestic actors could help to alleviate problems of implementation failure and non-compliance in the South.

The book is divided into three parts. Chapter two critically reviews the empirical evidence presented in the literature on the existence of a 'Southern Problem'. It starts with raising some methodological issues on the inferences that we can make from existing data. While the statistics on the infringement proceedings published by the European Commission in its Annual Reports help us to overcome the small-n problem of comparative case studies, the conclusions we can draw from the data on the level of member state compliance with European Law are limited. After having discussed some important problems with the empirical data available, the remainder of the chapter explores whether the European Union is facing a growing compliance problem and to what extent the Southern European member states can be held responsible. The analysis shows that we have no evidence for a widening implementation gap. And while non-compliance is more pronounced in the South than in the North, there is considerable variation between the member states, which cuts across the alleged North-South divide.

Chapter three revisits the various explanations offered by the literature on implementation failure and non-compliance with European Environmental Law. The first part of the chapter shows that neither the 'Mediterranean Syndrome' nor more general approaches to the 'Southern Problem' are able to account for the variations in member state compliance found in the first chapter. Nor are they able to explain policy variation within individual member states. The second part of the chapter develops an alternative model, which systematically links policy and country variables in explaining implementation and compliance. The 'Pull-and-Push' Model allows to reformulate the 'Southern Problem' in more general terms by identifying differing degrees of policy misfit and of pressure from below and from above as the major factors driving effective implementation and compliance.

Chapter four systematically tests the propositions of the 'Pull-and-Push' Model in a comparative study on the implementation of six different European environmental policies in Germany and Spain. It starts with some methodological issues on case selection and on measuring effective implementation and compliance. The 12 case studies differ with regard to the degree of policy misfit and the level domestic mobilization in the two countries. Germany represents an industrial and environmental first-comer with a high capacity to shape and implement European policies. The opposite is true for Spain, which belongs to the group of industrial and environmental late-comers. The two countries also differ with respect to the openness of their political opportunity structures and the mobilization capacities of their citizens and public interest groups. The main part of the chapter traces the processes through which the six policies have been legally implemented, practically applied, monitored, and enforced in Spain and Germany. The comparative study not only shows that environmental leaders may face substantial difficulties in complying with European environmental policies, too. It is the same factors that account for the compliance performance of the two member states.

The conclusions summarize the major findings of the book. Then, the scope conditions of the 'Pull-and-Push' Model are discussed. While its categories are sufficiently broad to be applied to other policy areas, it does not constitute a proper compliance theory. The model only identifies one albeit important causal mechanism of compliance, which can be applied across policy areas and various forms of law beyond the nation state. Finally, some of the implications of the 'Pull-and-Push' Model for European environmental policy-making are considered. Are we likely to see the rise of interest coalitions, which systematically pitch member states of diverse levels of socio-economic development against each other? Will eastern enlargement turn existent tensions between southern late-comers and northern first-comers into a true 'North-South conflict'? How can we improve compliance with European Environmental Law in an increasingly diverse European Union with more than 20 member states?

Chapter Two

Is there a 'Southern Problem'?

Member state compliance with European environmental policies is found to be wanting. Portugal, Italy, Greece, and Spain have the reputation of being particular laggards. But assessing member state compliance with European (Environmental) Law is not an easy task. This chapter reviews the evidence presented in the literature on the increasing implementation failure in European policy-making, for which the southern member states are often held responsible. It starts with raising some critical questions about the reliability of existing data. Drawing on some new sources, it then explores whether the implementation gap has been widening in the European Union and to what extent this can be attributed to a poor performance of its southern member states. The study finds only weak support for the overall perception of a 'Southern Problem'. Looking at different indicators, the picture is much more blurred and shows considerable variation both among the southern member states as well and the alleged 'North-South' divide.

Non-compliance in the European Union: Pathology or Statistical Artefact?[1]

For more than ten years, the European Commission has been denouncing a growing compliance deficit, which it believes to threaten both the effectiveness and the legitimacy of European policy-making (Commission of the European Communities 1990, 2000). While some scholars argue that the level of compliance with European Law compares well to the level of compliance with domestic law in democratic liberal states (Keohane and Hoffmann 1990: 278; Neyer, Wolf, and Zürn 1999), many consider non-compliance to be a serious problem of the EU that is systemic and pathological (Krislov, Ehlermann, and Weiler 1986; Weiler 1988; Snyder 1993; From and Stava 1993; Mendrinou 1996; Tallberg 1999). The contradicting assessments of member state compliance are partly explained by the absence of common assessment criteria and reliable data.

Most compliance and implementation studies develop their own assessment criteria and collect their empirical data in laborious field research (Knill 1998; Knill and Lenschow 1998; Duina 1997). As a result, a comparison of empirical findings and theoretical claims becomes difficult. Others therefore draw on statisti-

---

[1] A previous version of the following section has been published in the Journal of European Public Policy (Börzel 2000).

cal data published in the *Annual Reports on Monitoring the Application of Community Law* (Snyder 1993; Mendrinou 1996; Tallberg 1999; Macrory 1992; Collins and Earnshaw 1992; Pridham and Cini 1994). Article 226 (ex-Article 169) of the EC-Treaty entitles the Commission to open infringement proceedings against member states found in violation of European Law. Since 1984, the Commission has reported every year on the legal action it brought against the member states. There are five *types of infringements* of European Law against which the Commission may open proceedings (see figure 2.1):

1) *Violations of Treaty Provisions, Regulations, and Decisions ('violation')*
   Treaty Provisions, Regulations, and Decisions are directly applicable and, therefore, do not have to be incorporated into national law.[2] Non-compliance takes the form of not or incorrectly applying and enforcing European obligations as well as of taking, or not repealing, violative national measures.

2) *Non-transposition of Directives ('no measures notified')*
   Directives are not directly applicable, as a result of which they have to be incorporated into national law. Member states are left the choice as to the form and methods of implementation (within the doctrine of the *éffet utile*, which stipulates that the member states have to choose the most effective means).[3] Non-compliance manifests itself in a total failure to issue the required national legislation.

3) *Incorrect legal implementation of Directives ('not properly incorporated')*
   The transposition of Directives may be wrongful. Non-compliance takes the form of either incomplete or incorrect incorporation of Directives into national law. Parts of the obligations of the Directive are not enacted or national regulations deviate from European obligations because they are not amended and repealed, respectively.

4) *Improper application of Directive ('not properly applied')*
   Even if the legal implementation of a Directive is correct and complete, it still may not be practically applied. Non-compliance involves the active violation of taking conflicting national measures or the passive failure to invoke the obligations of the Directive. The latter also includes failures to effectively enforce European Law, that is to take positive action against violators, both by national administration and judicial organs, as well as to make adequate remedies available to the individual against infringements, which impinge on her rights.

---

[2]   Treaty Provisions and Regulations are generally binding and directly applicable, while Decisions are administrative acts aimed at specific individuals, companies, or governments for which they are binding.

[3]   ECJ Fédéchar v. High Authority, C-8/55; ECJ Van Gend en Loos, C-26/62.

5) *Non-compliance with ECJ judgements ('not yet complied with')*
Once the European Court of Justice finds a member state guilty of infringing European Law, the member state is ultimately obliged to remedy the issue. Non-compliance refers to the failure of member states to execute Court judgements, which establish a violation of European Law.

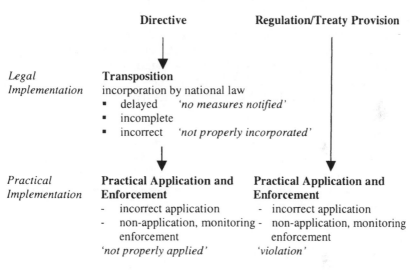

	Directive	**Regulation/Treaty Provision**
Legal Implementation	**Transposition** incorporation by national law ▪ delayed *'no measures notified'* ▪ incomplete ▪ incorrect *'not properly incorporated'*	
Practical Implementation	**Practical Application and Enforcement** - incorrect application - non-application, monitoring enforcement *'not properly applied'*	**Practical Application and Enforcement** - incorrect application - non-application, monitoring enforcement *'violation'*

Figure 2.1 Infringements of European Law

The proceedings specified in Article 226 consist of six subsequent *stages* (figure 2.2).

1) Suspected Infringement
Suspected infringements refer to instances in which the Commission has some reasons to believe that a member state violated European Law. Such suspicions can be triggered by different sources:
- *complaints* lodged by citizens, corporations, and non-governmental organizations;
- *own initiatives* of the Commission;
- *petitions* and *questions* by the European Parliament;
- *non-communication* of the transposition of Directives by the member states.

2) Formal Letter of Notice (Article 226)
The Formal Letter of the Commission delimits the subject matter and invites the member state to submit its observations. Member states have between one and two months time to respond. Unlike their name suggests, Formal Letters are not part of the official proceedings. The Commission considers them as a preliminary stage, which serves the purpose of information and consultation, affording a member state the opportunity to regularize its position rather than

bringing it to account (Commission of the European Communities 1984: 4-5).[4] Consequently, Formal Letters are only made official if they refer to cases where member states have not communicated the transposition of Directives within the given time limit. In such cases of non-transposition, the Commission automatically opens proceedings.

3) Reasoned Opinion (Article 226)
The Reasoned Opinion is the first official stage in the infringement proceedings. The Commission sets out the legal justification for commencing legal proceedings. It gives a detailed account of how it thinks European Law has been infringed by a member state and states a time limit, within which she expects the matter to be rectified. The member states have one month in which to respond.

4) Referral to the European Court of Justice (Article 226)
The ECJ Referral is the last means to which the Commission can resort in cases of persistent non-compliance. Before referring a case to the ECJ, the Commission usually attempts to find some last minute solutions in bilateral negotiations with the member state.

5) ECJ Judgement (Article 226)
The ECJ acts as the ultimate adjudicator between the Commission and the member states. First, it verifies whether a member state actually violated European Law as claimed by the Commission. Second, it examines whether the European legal act under consideration requires the measures demanded by the Commission. And finally, the Court decides whether to dismiss or grant the legal action of the Commission.

6) Post-Litigation Infringement Proceedings (Article 228)
If member states refuse to comply with an ECJ judgement, the Commission can open a new infringement proceeding for post-litigation non-compliance. Since 1996, it can ask the ECJ to impose financial penalties, either in form of a lump sum or a daily fine, which is calculated according to the scope and duration of the infringement as well as the capabilities of the member states.[5]

---

[4]   But note that, according to the view of the ECJ, the letter defines the object at issue in any subsequent court proceedings. As a result, the Commission is not allowed to include additional points during subsequent stages, even if she later discovers new infringements.

[5]   The basic amount of the fine is multiplied by a factor 'n', taking into account the GDP of a member state and its number of votes in the Council. The 'n' for Luxembourg, for instance, is 1 and for Germany 26.4 (OJ C 63, 28.2.1997).

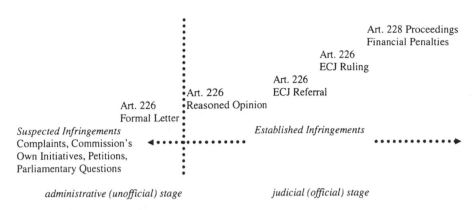

Figure 2.2   The different stages of the infringement proceedings

Various studies have used the numbers of infringements within the different stages as indicators for member state non-compliance with European Law. For instance, the observation that environmental policy accounts for over 20 per cent of registered infringement (Formal Letters) has been taken as evidence for a severe implementation deficit in this area (Commission of the European Communities 1996; Jordan 1999). Such inferences are not without problems though. There are good reasons to question whether infringement proceedings qualify as valid and reliable indicators of compliance failure, that is, whether they constitute a random sample of all the non-compliance cases that occur. First, for reasons of limited resources, the Commission is not capable of detecting and legally pursuing all instances of non-compliance with European Law. Second, for political reasons, the Commission may not disclose all the cases, in which it took action against infringements of European Law. Third, for methodological reasons, the infringement data are neither complete nor consistent.

The Problem of Undetected Non-compliance

Infringement proceedings only cover a fraction of the violations of European Law in the member states. The jurisprudence of the ECJ under Article 234 (ex-Article 177) already indicates that many cases of non-compliance occur without getting caught by the Article 226 procedure.[6] Infringement proceedings encompass cases of non-compliance, which have been detected by the Commission itself or have been brought to its attention by citizens, companies, and interest groups. The

---

6   According to the preliminary ruling procedure of Article 226, domestic courts may, and courts of last instance must, bring cases to the ECJ concerning questions of the legal interpretation of European Law. Those cases often arise if European regulations challenge national legal provisions.

detection rate is rather high for the failure to transpose Directives into national law. The Commission automatically opens proceedings after the transposition deadline of a Directive expired and a member state did not notify the Commission about its incorporation into national law. Non-transposition accounts for more than two-thirds of all infringement cases opened. The chances of detection significantly decrease when it comes to complete and correct transposition, practical application and enforcement of European policies. Due to its limited resources, the Commission largely depends on external sources, including member states reporting back on their implementation activities,[7] costly and time-consuming consultancy reports, or information from domestic actors. Commission officials can do on-site visits in the member states, but such spot-checks tend to be time-consuming, politically fraught, and can easily be blocked by member states. They are usually little more than 'fact-finding missions' to clarify certain points rather than investigate instances of suspected non-compliance.

Societal monitoring is the most important source of information for the Commission. Since the degree of social activism and respect for law varies among the member states, infringement proceedings may contain a serious bias. A country whose citizens are collectively active and law-abiding could generate more complaints than a member state whose citizens show little respect for the law and are less inclined to engage in collective action. Yet, the distribution of complaints across member states shows that societal activism per se is not the issue (table 2.1). Population size seems to be more important. The five biggest member states – Germany, France, the UK, Italy, and Spain – are home to more than 75 per cent of the European population and account for about 69 per cent of the complaints lodged between 1983 and 1999. At the same time, the numbers of complaints originating in Germany and the UK are lower than we would expect given their population size. Spain, by contrast, has an unusually high share of complaints compared to the other bigger member states. The same is true for Greece within the group of less populated states, since it accounts for a much bigger share of complaints than the Netherlands or Denmark. Both Spain and Greece show a lower degree of societal activism than their northern counterparts of similar population size. It has been argued that southern societies hold certain distrust against their state institutions as a result of which they resort to the European Union for assistance (Pridham and Cini 1994), which could explain the relatively high number of complaints originating in Spain and Greece. However, neither Italy nor Portugal fit this explanation, since their societies may be less active but their numbers of complaints are relatively low compared to their population size.

---

[7] Only Denmark, Finland, and Sweden regularly report to the Commissions the measures taken to transpose EU Directives into national law (Jordan 1999).

Table 2.1[8] **Member states compared by population, complaints and infringement proceedings opened, 1983-99**[9]

	Percentage of EU population	Average percentage of complaints	Average percentage of proceedings opened
Germany	21.9%	11.9%	7.8%
France	15.7%	16.8%	10.3%
UK	15.7%	9.9%	6.6%
Italy	15.3%	12.9%	11.6%
Spain	10.6%	17.6%	10.1%
Netherlands	4.2%	3.5%	5.9%
Greece	2.8%	10.5%	11.3%
Belgium	2.7%	5.1%	8.4%
Portugal	2.6%	4.5%	10.8%
Denmark	1.4%	2.6%	4.5%
Ireland	1.0%	3.8%	6.5%
Luxembourg	0.1%	0.9%	6.2%

Source: column 1: OECD compendium; column 2, 3: Annual Reports on the Monitoring of the Application of European Law 1984-99: www.iue.it/RSCAS/Research/Tools/.

Another factor, which could bias the detection of non-compliance with European Law, is linked to the availability of reliable data. Some member states may lack the necessary administrative capacity to verify whether European legislation is complied with. Monitoring water and air quality, for instance, requires an adequate technical and scientific infrastructure. In the absence of comprehensive and reliable monitoring data, neither the member states nor their citizens nor the Commission

---

[8] In order to compare the member states, which differ in their years of membership, I standardized their scores. First, I divided the number of complaints, letters etc. of the different member states by their years of membership. Second, I added up these average scores and made the sum equal 100 per cent. Finally, I calculated the percentage of the average scores.

[9] Finland, Austria, and Sweden are excluded because they joined the EU only in 1995. They are still in the adaptation phase and the incorporation of the comprehensive *acquis communautaire* into national law is not fully concluded yet. Most of their infringement cases refer to the delayed transposition of Directives. Therefore, their infringement records are likely to be exaggerated, particularly in the earlier stages of the proceedings.

are able to assess compliance with European air and water pollution control Directives. Yet, member states with high monitoring capacities, such as Denmark and the Netherlands, show a low number of complaints and infringement proceedings opened while those with weaker administrative and scientific infrastructures, like Greece and Spain, find themselves at the upper end of the list (table 2.1). Moreover, it has been argued in the literature that it is the very lack of monitoring capacity in some (southern) member states, which, among other factors, accounts for their high number of infringements (Pridham and Cini 1994; Hooghe 1993).

In sum, we have no indication that the limited detection of non-compliance would systematically bias the infringement data.

The Problem of Selective Disclosure of Detected Non-compliance

The Commission has considerable discretion in deciding whether and when to open official proceedings (Evans 1979; Audretsch 1986). In principle, the Commission prefers informal bargaining to formal sanctions in order to induce member state compliance (Snyder 1993). It considers an official opening of Article 226 proceedings only 'when all other means have failed' (Commission of the European Communities 1991: 205). The great majority of cases are settled in bilateral exchanges with national authorities during the administrative stage – only about one third of the letters result into Reasoned Opinions and, hence, become official. The sending of a Formal Letter is already preceded by written exchanges and meetings between the Commission and the member state on an informal level. The political discretion of the Commission in deciding whether and when to open official proceedings could cause a voluntary bias in the sample. This might be all the more true since the Article 130r(4) of the Treaty attributes the primary responsibility for implementing EU policies to the member states. The principle of decentralized enforcement of European Law puts the Commission, which does not enjoy any direct political legitimacy, in a weak and 'invidious position' (Williams 1994). Thus, the Commission may treat some member states more carefully than others because they make significant contributions to the EU budget or dispose of considerable voting power in the Council. Or their population tends to be 'Eurosceptic' and the Commission seeks to avoid to upset the public opinion in these member states by officially shaming them for non-compliance with European Law (Jordan 1999).

The comparison between the relative ranking of the member states at the unofficial (Formal Letters) and the first official stage (Reasoned Opinions) of the proceedings could help us to reveal such a bias (table 2.2). Germany and France are the two member states, which contribute most to the EU budget and possess considerable bargaining power in the Council. Nevertheless, they both figure prominently among those member states that have received high numbers of Reasoned Opinions. While those two countries are rather pro-European, public and elite support for European institutions in Denmark and the UK is among the lowest, only topped by Austria and Sweden, which recently joined the European Union.[10] Denmark does indeed perform best among the member states at both stages. The British record, however, is more mixed.

Table 2.2 Ranking of member states at the stages of Formal Letters and Reasoned Opinions, 1978-99

Formal Letters	Reasoned Opinion	
Italy	Italy	high level of non-compliance
Greece	Greece	
Portugal	Portugal	
France	France	
Spain	Belgium	
Belgium	Spain	
Germany	Germany	
Ireland	Ireland	
UK	Luxembourg	
Luxembourg	UK	
Netherlands	Netherlands	
Denmark	Denmark	low level of non-compliance

Source: Annual Reports on the Monitoring of the Application of European Law 1984-99: www.iue.it/RSCAS/Research/Tools/.

In sum, there are no obvious factors that bias our sample towards politically less sensitive cases and member states, respectively. This should not be too surprising. The Commission may have considerable discretion in taking action against member state non-compliance. But there is a powerful organizational 'logic of appropriateness' (March and Olsen 1998) that prevents the Commission from abusing its discretion. The Commission's authority as the 'Guardian of the Treaties' first of all depends on its credibility as an impartial adjudicator between competing interests. Its identity as a truly supranational body makes it inappropriate for Commissioners and Commission officials to block legal action against their own member state when it stands accused of violating against European Law (Egeberg 2001: 739).

---

[10] See 'Initial Results of Eurobarometer Survey No. 54 (autumn 2000)', Brussels: European Union, 8 February 2001.

Not only do the member states seek to make sure that they are not treated in an unfair way in comparison to the others. For infringement decisions are ultimately taken in the College, there is a certain amount of self-control within the Commission itself. Thus, President Prodi expressed his intention to shame Commissioners who bow to domestic pressure favouring some member states over others.[11]

The Problem of Incomplete and Inconsistent Data

The infringement data published by the Commission are neither complete nor consistent. First, the Commission has repeatedly changed the way in which it reports data. Suspected infringements are a case in point. From 1982 till 1991, their numbers are indicated by two different figures, complaints and own investigations by the Commission. Between 1992 and 1997, the Commission provides only one figure, which does neither refer to complaints only nor to the Commission's own investigation nor does it equal the aggregate of the two. Since 1998, the Commission reports three figures – complaints, own investigations, and non-communication of the transposition of Directives, whereby it remains unclear whether the third category has been newly introduced or used to be an integral part of one of the other two categories.

A comparison of suspected infringements across time is further impaired because since 1995, the Commission has subsumed parliamentary questions and petitions under complaints and own investigations, respectively. Additional problems arise when it comes to the reporting of infringement cases by policy sectors, since the Commission has redefined them several times over the years. In 1992, for instance, the Directorate General III changed its name from 'Internal Market and Industrial Affairs' to 'Industry' as a result of which the number of complaints in this sector dropped dramatically from 382 in 1992 to 34 in 1993. Finally, some data are not provided at all or only for a limited number of years. Transposition rates for Directives have been included in the Annual Reports as late as 1990. Since 1998, figures for suspected infringements are merely given by member states, unlike in previous years, where they were also provided by policy sector. Established infringements, finally, are jointly reported by policy sector *and* member states only in the 10th Annual Report for the years 1988 till 1992 (Commission of the European Communities 1993b: 165ff.). In 1992, the Commission also stopped reporting Court Judgements.

Second, the reported data show some serious inconsistencies. For any given year, the Annual Reports of the Commission provide two types of data. Aggregate data summarize the number of infringement proceedings classified by the different stages, member states, policy sectors, and type of infringements. The 'raw' data list the individual infringement cases, which are to make up the aggregate data. The comparison of the aggregate and the raw data reveals some serious 'mismatches'.

---

[11]   *European Voice*, 30 September–6 October 1999.

The raw data merely comprise about one third of the Letters actually sent. This is explained by the policy of the Commission to individually list Letters only if they refer to cases of non-transposition.[12] But the aggregate data for Reasoned Opinions and Court Referrals do not equal the sum of the individually listed cases either. The aggregate data report 5762 Reasoned Opinions sent by the Commission between 1978 and 1999. But the 17 Annual Reports (1984-1999) list only 4241 Reasoned Opinions for these years; some 26.4 per cent of the cases are missing. The same inconsistencies can be found for ECJ Referrals, where about 37.9 per cent of the cases are not listed (1593 to 990).

Confronted with the inconsistencies in their published data, the Commission provided the author with a dataset drawn from its own database containing all cases of Reasoned Opinions, Court Referrals, and Court Judgements. Their aggregate numbers closely correspond to the aggregate data published in the Annual Reports (table 2.3).

Table 2.3   Comparing infringement data from the Annual Reports and the Commission, 1978-99

	Reports – Individual Listings	Reports – Aggregate Date	Dataset provided by Commission
Reasoned Opinion	4243	5762	5760
Court Referrals	993	1593	1618
Court Judgements	488	430*	672

Source: Annual Reports on the Monitoring of the Application of European Law 1984-99, www.iue.it/RSCAS/Research/Tools/.
* The Annual Reports provide data only for 1978-92.

The explanation for the poor goodness of fit between the published aggregate data and the published raw data lies in the reporting methods. Unlike in the aggregate data, only those cases are individually listed that are still open at the end of the year reported. For instance, if the Commission had sent a Reasoned Opinion in January and the case is closed in July because the member state rectified the violation, the case features in the aggregate but not in the raw data. In 1999, 122 out of 438 cases, in which the Commission had sent a Reasoned Opinion, were closed or merged with similar cases. Most of them (104) refer to the delayed transposition of

---

[12]   The reports do list a few hundred other Letters because, for political reasons, the Commission sometimes decides to make a Letter public. Moreover, some Directorates General are less faithful to the Commission's policy of not disclosing cases of improper incorporation and application.

Directives.[13] In sum, the incompleteness and inconsistencies in the published infringement data appear to be the result of changing reporting methods rather than administrative 'sloppiness' or political manipulation.

In conclusion, the Commission data on member state infringements of European Law suffer from some problems, which should caution us against their use as straightforward indicators of non-compliance with European Law. At the same time, the Commission data are the only statistical source available. Neither international organizations nor states provide such comprehensive information on issues of non-compliance. The Infringement database compiled by the author comprises some 6230 infringements cases, which the Commission initiated between 1978 and 1999. Since the Commission does not fully report Formal Letters, the database only contains the individually listed cases of Reasoned Opinions and subsequent stages. The cases are classified by infringement number, member state, policy sector, legal basis (celex number), legal act, type of infringement, and measures taken by the Commission. Unlike the aggregate data in the Annual Reports, each case features only once and not several times as a Letter, Reasoned Opinion, ECJ Referral etc. These data can serve as important indicators for non-compliance as long as we carefully control for potential selection biases.

But even if we accept infringement data as valid and reliable indicators of member state non-compliance with European Law, we have to be careful in how to interpret them. It is a commonly held assumption – both among policy makers and academics – that the EU is facing a growing compliance problem that is systematic and pathological. The negative assessment is backed by the increasing number of infringement proceedings (Formal Letters), which the Commission has opened against the member states over the years (see chart 2.1).

Since 1978, the Commission has opened more than 17,000 infringement proceedings against the member states. This figure may sound impressive but must be put into perspective. Infringement numbers as such do not tell much about either the absolute scope of non-compliance or relative changes in the level of non-compliance over time. Infringement cases only cover a fraction of member state violations against European Law. We may claim that they provide a representative sample, but we have no means to estimate the total number of the population of non-compliance cases.

---

[13] Interview in the enforcement unit of the Secretariat General of the Commission, Brussels, 04/01.

Chart 2.1 Total number of infringement proceedings opened for the EC 12, 1978-99

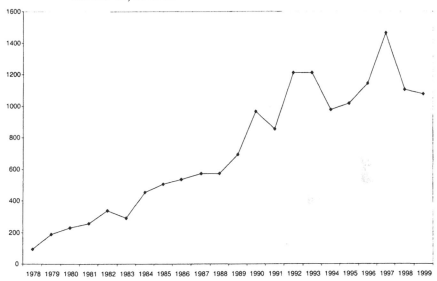

Source: Commission of the European Communities 2000.

The available data do not permit us to draw any inferences about the existence or non-existence of a compliance problem in the European Union. We can only trace relative changes in non-compliance, that is, assess whether non-compliance has increased or decreased over time. But in order to do this, we have to measure the number of infringement proceedings opened against the numbers of legal acts that can be potentially infringed as well as the number of member states that can potentially infringe them. Between 1983 and 1998, the number of legal acts in force has more than doubled (from 4566 to 9767)[14] and five more member states have joined the Union. If we calculate the number of infringement proceedings opened as a percentage of 'violative opportunities'[15] (number of legal acts in force multiplied by member states) for each year, the level of non-compliance has not increased. This is particularly true if we control for several statistical artefacts that inflate the infringement numbers. First, the Commission adopted a more rigorous approach to member state non-compliance in the late 1970s (Mendrinou 1996: 3). Likewise, the Commission and the ECJ pursued a more aggressive enforcement policy in the early 1990s in order to ensure the effective implementation of the Internal Market Programme (Tallberg 1999). Not surprisingly, the numbers of opened infringement

---

[14] I am thankful to Wolfgang Wessels and Andreas Maurer for providing me with the annual numbers of legislation in force.

[15] I owe this term to Beth Simmons.

proceedings increased dramatically twice, in 1983-84 by 57 per cent and again in 1991-92 by 40 per cent. Second, the Southern enlargement in the first half of the 1980s (Greece, 1981, Spain and Portugal, 1986) led to a significant increase in infringement proceedings opened once the 'period of grace', which the Commission grants to new member states, had elapsed. From 1989 to 1990, the number of opened proceedings grew by 40 per cent (223 cases), for which Spain, Portugal, and Greece are single-handedly responsible. The three countries account for 249 new cases while the numbers for the other member states remained more or less stable. The last significant increase of 28 per cent in 1996-97, finally, is not so much caused by the Northern enlargement (Sweden, Austria, Finland 1995) but by a policy change of the Commission. In 1996, the internal reform of the infringement proceedings re-stated the 'intended meaning' (*sense véritable*) of the Formal Letters as mere 'requests for observations' (*demande d'observation*) rather than warnings of the Commission.[16] Avoiding any accusations, Letters should be issued more rapidly than before. Indeed, the number of Letters sent grew significantly after the reform had been implemented. If all these factors are taken into account, the number of infringements has not significantly increased over the years but remained rather stable (chart 2.2).

Chart 2.2 Total number of infringement proceedings opened in relation to violative opportunities for the EC 12, 1983-98

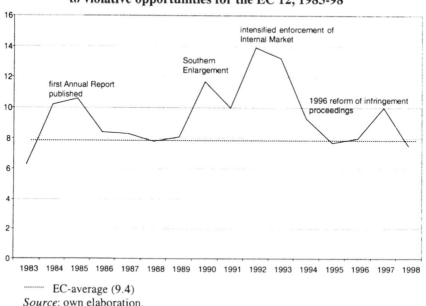

········ EC-average (9.4)
Source: own elaboration.

---

16   Internal document of the Commission, unpublished.

To sum up, the infringement data do not allow us to make any statements about the absolute level of non-compliance in the European Union. We can use the data, however, for comparing relative levels of non-compliance across time, policy sectors, and member states.

Exploring the 'Southern Problem'

Non-compliance with European Environmental Law is rather high relative to other policy areas (see table 2.4). The average transposition rate of environmental Directives (91.9 per cent) is satisfactory and close to the Community average (92.5 per cent). But about a quarter of all complaints received by the Commission refer to the environment. Environmental infringements account for 14.6 per cent of all Reasoned Opinions sent between 1978 and 1999, for 17.5 per cent of the cases referred to the ECJ and for 19.8 per cent of the ECJ Rulings in the same time period. Finally, half of the cases that the Commission brought before the ECJ for non-compliance with its rulings, fall into the area of the environment.

Table 2.4   Percentage of environmental infringements by stages and share of southern member states, 1978-99[17]

	Complaints (1982-97)	Letters	Reasoned Opinions	ECJ Referrals	ECJ Judgements	Art. 228 ECJ Referrals
Environment (percentage of total infringements)	24.8%	Data available only for 1988-92	14.6%	17.5%	19.8%	4 out of 8 cases
Italy, Greece, Spain, Portugal (percentage of environmental infringements)	42.2%		43.3%	41.8%	35.3%	1 out of 4 (Greece)

Source: column 1: Annual Reports; column 3-6: EUI Database on Member State Compliance with Community Law: www.iue.it/RSCAS/Research/Tools/.

The relative compliance record of the environment looks even worse, if we consider that environmental legislation only accounts for roughly three per cent of the legislation in force. Although we have no data on whether the implementation gap

---

[17]   Finland, Austria, and Sweden, which joined the EU in 1995, are excluded from the table because they might be over represented in the earlier stages (see above) and underrepresented in the higher stages.

has been widening, non-compliance with European Law is certainly more prevalent in the area of environment than in other policy sectors.

But to what extent is non-compliance with European Environmental Law a
'Southern Problem'? Which share do Portugal, Italy, Greece, and Spain have in the
infringements of European Environmental Law? The data do not lend themselves
to an easy conclusion (table 2.4). Taken together, the four Southern European
member states account for about 40 per cent of environmental Reasoned Opinions
and ECJ Referrals. But their share drops once the cases reach the ECJ, approaching
the 33 per cent mark, which they would hit if the infringements were equally
distributed among the 12 member states. Moreover, four northern countries
(Germany, France, Belgium, and Luxembourg) account for one third of the
Reasoned Opinions, 44.6 per cent of the Court Referrals, and 57.1 per cent of the
Court Rulings. Finally, the performance of the four southern countries varies
considerably for the different compliance indicators, including transposition rates
of Directives and the number of infringement cases across the various stages of the
infringement proceedings.

Member state transposition of environmental Directives, which make up for
over 95 per cent of European environmental legislation, is, in general, quite satisfactory. Their incorporation into national law is usually only a matter of time. In
most cases, member states are just not able to meet the deadlines set by the Directives but enact the necessary legislation with some delay.[18] Unlike Spain and Portugal, whose transposition records compare well with several northern countries,
Italy and Greece showed a rather poor performance in the early 1990s. But in the
second half of the 1990s, their records vastly improved and closed in with the other
member states (see table 2.5). The range between the transposition rates of the
member states used to be greatest in 1991, with 39 per cent, which is explained by
the Single Market Programme imposing a heavy implementation burden on the
member states. In 1999, however, the range shrunk to nine per cent, with all member states scoring better than 90 per cent. The average range for the EC 12 between
1990 and 1999 is only 16 per cent, with Denmark being the top leader (99 per cent)
and Italy lagging behind (83 per cent). The two opposite ends of the list fit the
North-South dichotomy. But in-between we find variation that does not follow this
pattern. The average transposition rate of Portugal and Spain is around EC average
(92 per cent), and compares well against Germany, France, the UK, Luxembourg,
Ireland. Belgium, by contrast, remains below EC average as Greece does.

---

[18]   Whether transposition is correct or complete, is, of course, an entirely different matter.

Table 2.5   Transposition rates by member states for the environment, 1990-99

	BELG	DK	GER	Greece	Spain	France	Ireland	Italy	Lux	NL	Port.	UK	EC 12
1990	86%	99%	92%	79%	92%	92%	87%	63%	89%	97%	95%	91%	89%
1991	81%	98%	92%	76%	93%	89%	84%	59%	86%	95%	94%	85%	86%
1992	94%	99%	92%	92%	91%	96%	90%	83%	92%	97%	89%	93%	92%
1993	91%	98%	91%	84%	90%	95%	88%	81%	92%	92%	90%	90%	90%
1994	85%	100%	91%	85%	86%	94%	97%	76%	93%	98%	82%	82%	89%
1995	83%	98%	94%	88%	90%	95%	95%	85%	92%	98%	87%	93%	92%
1996	86%	98%	96%	91%	94%	93%	96%	85%	96%	98%	94%	94%	93%
1997	87%	100%	94%	97%	99%	96%	98%	97%	98%	99%	97%	96%	97%
1998	96%	99%	98%	95%	98%	97%	98%	97%	99%	100%	97%	96%	98%
1999	92%	99%	91%	91%	93%	92%	97%	99%	92%	100%	97%	91%	95%
EC12	88%	99%	93%	88%	93%	94%	93%	83%	93%	97%	92%	91%	92%

Source: Annual Reports 1990-99.

While cross-national differences in transposition of European Directives have levelled, member state performance still varies significantly between the stages of the infringement proceedings, also across the North-South divide (table 2.6).

Table 2.6   Member state non-compliance with European Environmental Law, 1978-99

	Transposition Rates (1990-99)	Formal Letters (n.a.)	Reasoned Opinions (1978-99)	226 ECJ Referrals (1978-99)	226 ECJ Rulings (1978-99)	228 ECJ Referrals (1978-99)
Italy	82.5%		13.1%	14.2%	17.0%	0
Greece	87.8%		12.6%	14.2%	6.6%	1
Spain	92.6%		10.3%	8.1%	8.9%	0
Portugal	92.2%		13.2%	11.2%	6.7%	0
France	93.9%		7.7%	8.4%	8.5%	1
Belgium	88.1%		9.9%	14.5%	22.7%	0
Germany	93.1%		7.7%	10.3%	17.0%	2
Ireland	82.5%		8.1%	6.7%	0.7%	0
UK	91.1%		6.0%	2.3%	2.1%	0
Luxemb.	92.9%		6.2%	8.1%	6.4%	0
Netherl.	97.4%		3.7%	1.6%	3.5%	0
Denmark	98.8%		1.6%	0.3%	0%	0

Source: column 1, 2: Annual Reports; column 3-6: EUI Database on Member State Compliance with Community Law: www.iue.it/RSCAS/Research/Tools/.

Italy is the only southern member state that shows a consistently bad performance across all stages. The initial records of Greece, and Portugal are equally poor but show an improving trend towards the later stages of the infringement proceedings where they are passed by some of the northern countries, notably Belgium and Germany. Germany is usually considered as one of the environmental leaders or pioneers (Andersen and Liefferink 1997). But its judicial compliance record looks increasingly more like the one of an environmental laggard; it scores fifth on ECJ referrals and second on ECJ Rulings, together with Italy. Moreover, it holds the price for post-litigation non-compliance – two out of the four ECJ Referrals for non-compliance with a previous ECJ Ruling concern Germany (France and Greece account for the other two). Spain shows a medium level of non-compliance, which only closes in with the other three Southerners at the stage of ECJ Rulings.

The member states not only differ in their levels of non-compliance. They also show varying patterns of non-compliance (chart 2.3). Denmark, the Netherlands, and the UK maintain a rather 'decent' level of non-compliance, which ranges well below the Community average. Luxembourg, France and Ireland also stay within average and below, while the Irish non-compliance record declines toward the later stages approaching those of the three compliance leaders. By contrast, the initial level of non-compliance of Greece, Portugal, and Spain is above average but improves when reaching the ECJ (for Spain to a lesser extent though). The opposite is true for Italy, Germany, and Belgium whose record is deteriorating with each stage.

Chart 2.3   Non-compliance across stages for the EC 12, 1978-99

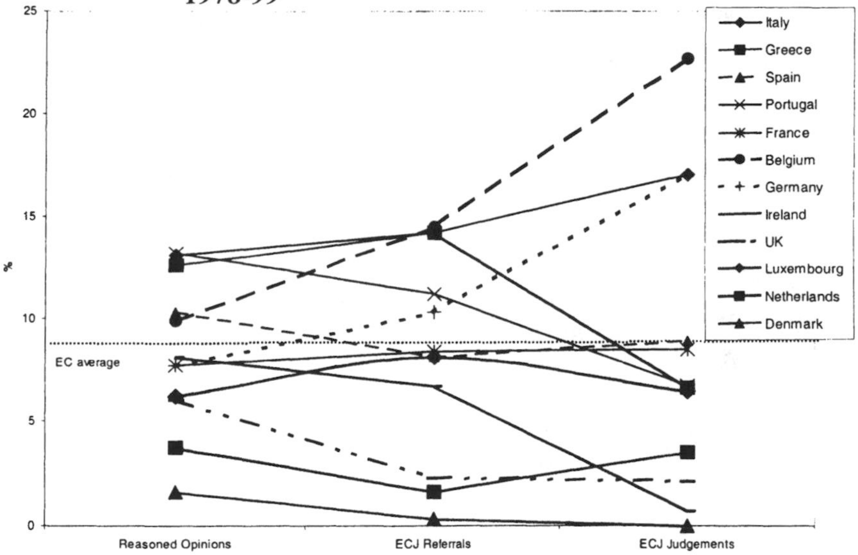

Source: table 2.6.

In sum, the compliance record of the member states is rather differential and does not fit a clear North-South pattern. The model pupils certainly come from the North and are led by Denmark and the Netherlands. But the slovenly fellows are only partly found in the South. Italy tops the list of compliance laggards where we also find Greece and Portugal. But the group is closely followed by Belgium and Germany, which rank considerably higher than Greece, Portugal and Spain in the later stages of the proceedings. These findings may be counterintuitive, particularly for Portugal and Greece on the one hand, and Germany, on the other hand, since they almost seem to switch roles when reaching the judicial stages. There are several implementation studies, which confirm that Germany's compliance record does not match its reputation as an environmental leader (Knill and Lenschow 1998; see also chapter 4). Portugal and Greece, by contrast, are still rather unknown territory. The literature offers little information, which would allow us to verify Portugal's positive performance reflected in the low number of official infringement proceedings. Part of Portugal's success has been credited to the massive infusion of capital into its infrastructure, of which Community funds made up 75 per cent of the costs (OECD 1994: 116-117). But there are hardly any empirical studies that systematically evaluate the implementation of European environmental policies in this country.

We may reject the above findings altogether arguing that infringement proceedings are not a valid indicator of non-compliance with European Environmental Law. But several studies have used these data to support arguments about the existence of a 'Southern Problem' in the implementation of European environmental policies (La Spina and Sciortino 1993; Pridham and Cini 1994; Aguilar Fernandez 1994). The following chapter presents a comparative study on the implementation of six European environmental Directives in Germany and Spain, which will further challenge this North-South dichotomy. The findings will substantiate the claim that northern and southern countries may equally face problems in complying with European Environmental Law. Beside the question whether there is a 'Southern Problem' and how pronounced it may be, the major challenge is to develop a theoretical approach that is general enough to account for variation among the different member states – a variation which clearly cuts across the North-South divide.

Chapter Three

Why there is (not) a 'Southern Problem'

The documented non-compliance record of the member states may not follow a North-South pattern. But the four southern member states do face significant compliance problems – as some of the northern member states do. This chapter discusses various explanations offered by the literature on the implementation of European environmental policy and scrutinizes them for their capacity to account for the varying degree of compliance in the four southern member states. Particular attention is given to the so called 'Mediterranean Syndrome' approach, which argues that the Southerners share some genuine defects in their culture that make it difficult, if not impossible, for them to comply with European Environmental Law. The first part of the chapter argues that neither the 'Mediterranean Syndrome' nor more general approaches can explain the varying compliance records of the (southern) member states. Nor can they account for the overall variation observed across the North-South divide. The second part develops an alternative model, which systematically links European and domestic factors in explaining compliance and implementation. The model is not only able to account for cross-country variation in the various parts of Europe. It helps to understand why the member states comply better with some policies than with others.

The Diagnosis is the 'Mediterranean Syndrome'

The literature which considers non-compliance and implementation failure in EU environmental policy a peculiar 'Southern Problem' argues that the Southern European member states have developed a civic culture which impedes the successful provision of public goods, such as a clean environment. The 'Mediterranean Syndrome' approach was most systematically developed in a controversial article co-authored by two Italians, Antonio La Spina and Giuseppe Sciortino (La Spina and Sciortino 1993). It resonates, however, with other works on (the absence of) social trust in southern societies (Banfield 1958; Putnam 1993). For socio-historical factors, which have been largely absent in the northern parts of Europe, Mediterranean countries are held unable to develop an organizational structure capable of promoting collective action. The 'Mediterranean Syndrome' summarizes the three main deficiencies that account for this incapacity (cf. La Spina and Sciortino 1993: 219-222). First, a civic culture, which is characterized by 'amoral familism' (Banfield 1958) and is void of social capital, that is, norms and social relationships which sanction cooperative and compliant behaviour (Putnam 1993). Citizens are not concerned with public affairs. They abstain from voluntary work or public actions, and do not hold public officials accountable. Nor do they voluntarily comply with the law. Law-abidingness strongly depends on the fear of punishment. Public officials are little concerned with the public interest either. They abuse their office to their private advantage. Bribery and corruption are widespread. As a result, social organization is highly segmented and ridden by distrust rendering collective action difficult and unstable.

Second, administrative structures and traditions make enforcement of public policies troublesome and random. Public administration is fragmented. Responsibilities are dispersed across different sectors and levels of government without providing for effective coordination mechanisms and sufficient technical expertise. Subservience to fascist regimes has undermined the legitimacy of the bureaucracy. Civil service enjoys little social prestige. Corruption is complemented by clientelism because civil servants depend on politicians for recruitment and promotion. While citizens are hardly inclined to comply with the law, public authorities do little to enforce it.

Third, a fragmented, reactive, and party-dominated legislative process impedes the enactment of effective public goods regulations. Parliamentary government coupled with proportional representation prevents the emergence of an executive with strong regulatory powers. Moreover, in the absence of powerful social movements, parties dominate the legislative process. Their proportional representation in Parliament renders decision-making cumbersome and difficult. Coalition governments tend to avoid controversial issues and postpone politically inconvenient decisions. Whereas public goods regulations stand little chance to be effectively implemented and complied with, they often do not come into existence in the first place.

In sum, if the conditions of the 'Mediterranean Syndrome' hold, the provision of public goods becomes highly problematic. Citizens 'stubbornly free-ride' while policy-makers and public authorities are neither willing nor able to overcome the impasse by enacting public goods regulations and enforcing them (La Spina and Sciortino 1993: 220). Its protagonists are quick to admit that the 'Mediterranean Syndrome' (MS) presents more a caricature than a portrait of reality. The purpose of the exercise is to 'create an ideal type' which 'allows some gain at the theoretical level' (La Spina and Sciortino 1993: 231, Fn. 4). The MS depicts a 'structural tendency' rather than a reality in the four Mediterranean member states (ibid: 221). None of them suffers from all the symptoms. Indeed, Italy and Greece seem to be more stricken than Spain and Portugal, which could account for the variation in their environmental performance. But the authors still maintain that the four countries satisfy the three general conditions of the MS: '1) dispersed micro-behaviour of citizens which is difficult to monitor, 2) complex administrative procedures, requiring a high level of technical expertise, and 3) a contradiction between high legitimation of the policy goal and the potential costs of effective decision-making' (La Spina and Sciortino 1993: 221, ad verbatim quote). Since all four Southern European countries show these symptoms, their chances of effectively implementing and complying with European Environmental Law are doomed.

The conditions of the MS are not well specified, which makes it hard to empirically falsify them. How do we observe, for instance, dispersed micro-behaviour of citizens? Robert Putnam has successfully failed in his attempt to measure the 'civicness' of regional societies in Italy (Putnam 1993). But even if we were able to operationalize the 'Mediterranean Syndrome', we would still find that the four southern countries are too diverse to be subsumed under one homogeneous phenomenon. Moreover, some of the MS symptoms are not confined to the South but are equally present in the North.[1] Public administration, for instance, is rather centralized in Greece and Portugal. Spain and Italy, by contrast, devolved considerable competencies to the regional level. The complexity of their administrative structures can easily compete with federal states, like Belgium or Germany, of which the latter also underwent a fascist regime. While all four Southern European countries are parliamentary systems based on proportional representation,[2] only Italy has experienced shifting majorities and short-lived governments. Spain, Portugal, and Greece prove to be politically as stable as Northern European countries with coalition governments, like Germany, the Netherlands or Denmark. Corruption and clientelism, finally, are certainly more prevalent in the South than in the North. Yet, Italy and Greece range much higher on corruption indexes than Spain and Portugal, which are closer to the ratings of Belgium and France (Treisman 2000). While some Mediterranean regions are still plagued by clientelism, notably southern Italy (Grote 1997), others, like Emiglia Romagna or Tuscany, have long developed alternative relations of social exchange.

Only Italy, and to some extent Greece, seem to fit the diagnosis of the 'Mediterranean Syndrome'. The two countries do indeed head the list of compliance laggards. Thus, some of the MS symptoms may have a causal effect on non-compliance. But instead of renaming the 'Mediterranean Syndrome' into the 'Roman-Hellenic Syndrome', we should seek to identify the causal factors which are behind the symptoms in order to arrive at explanations that are not reductionist and geographically bound.

Beyond the 'Mediterranean Syndrome'

Some authors have tried to overcome the reductionism of the 'Mediterranean Syndrome' (Pridham 1994; Pridham 1996; Pridham and Cini 1994; Aguilar Fernandez 1994; Aguilar Fernández 1997; Yearley, Baker, and Milton 1994). In order to

---

[1] La Spina and Sciortino define the Mediterranean countries as Spain, Italy, Greece, and Portugal equating the Mediterranean region with Southern Europe. They conveniently overlook that Portugal does not border the Mediterranean Sea while they exclude France, which does.

[2] The Portuguese president is directly elected but has only weak powers. Portugal operates much like a parliamentary system.

account for the problems of southern countries in complying with European Environmental Law, they adopt a more universal approach drawing on explanatory factors, which are present in all member states but are found to be more pronounced in the South. Ultimately, they come to similar conclusions than the proponents of the 'Mediterranean Syndrome' without, however, fully sharing their fatalistic and reductionist impetus.

Alternative approaches to the 'Mediterranean Syndrome' start with the observation that problems of implementation failure and non-compliance are prevalent but not confined to the southern member states. They identify a wide range of explanations for poor implementation that can be organized into four groups of variables rooted in the domestic structures of the member state (cf. Pridham 1996; Rehbinder and Stewart 1985):

- political variables referring to the degree to which environmental competencies are dispersed between different policy sectors and between different levels of government, on the one hand, and to the absence or presence of mechanisms of horizontal and vertical coordination, on the other hand;
- administrative variables related to administrative style and administrative resources in environmental policy-making;
- economic variables concerning the level of socio-economic development, which affects the costs of applying EU environmental legislation as well as the capacity and willingness of actors to cope with them;
- cultural variables regarding public awareness and the readiness to engage in collective action.

The variables are general enough to be applied to all member states but are found to vary significantly across the North-South divide. First, the four southern countries are said to suffer more from horizontal and vertical fragmentation of their administrative structures than their northern counterparts (Pridham and Cini 1994: 264-268; Pridham 1994: 84-85, 1996: 52; Yearley, Baker, and Milton 1994: 15). Environmental functions may be equally dispersed but, unlike the North, the South lacks effective coordination mechanisms. In all four southern member states, various national ministries hold environmental competencies, among which the Ministry of Environment has little weight. Ministerial rivalry and bureaucratic lethargy prevent effective inter-ministerial coordination. Horizontal fragmentation is complemented by vertical fragmentation in Spain and Italy, where the regions hold considerable responsibilities in environmental policy. Weak coordination between national and regional administrations has given rise to uneven implementation and variation in policy performance.

Second, the Southerners lack the administrative capacity to effectively make and implement environmental policies. Policy-making in these countries is *ad hoc*, reactive, and non-cooperative. Environmental protection enjoys low political priority and is often seen in an antagonistic relationship with employment and economic growth. Traditionally, policies have come about in response to crisis or emergency rather than as a matter of prevention and environmental management strategies (Pridham 1994: 86-88; Weale et al. 2000: chapter 6). The absence of institutionalized forms of consultation and cooperation with societal and economic actors further prevents the evolution of new, more proactive policy initiatives. The reactive and closed policy style often contradicts the proactive approach embodied in EU environmental policies (Aguilar Fernandez 1994; Pridham 1994: 88-90, 1996: 53). Moreover, southern administrations often do not possess sufficient technical expertise, staff, and infrastructure to effectively apply and enforce EU environmental legislation (Commission of the European Communities 1991: 213; Pridham 1994: 89-90). Data collection and regular information on the state of the environment are not efficient. Monitoring systems are incomplete and not always reliable. Insufficient funding for adequate measurement technologies and trained personnel to apply them enhances the problems of effectively controlling compliance with European environmental standards. Finally, the absence of administrative review and outcome control provides little incentive for administrators to enforce European environmental standards against recalcitrant private actors and causes a major discrepancy between formal and practical implementation of European policies.

Third, Southern European states have a lower level of socio-economic development which imposes serious funding constraints and renders the implementation of European environmental policies often politically prohibitive due to powerful economic interests and the need to preserve and create employment (Pridham and Cini 1994: 255-256; Kousis 1994). Not only business interests but also trade unions and the wider public tend to perceive environmental protection as a threat to creating and securing employment and economic growth. The promotion of economic development remains a priority of both public and private actors (Weale et al. 2000: chapter 5).

Finally, political activism and environmental awareness are only emerging in Southern European societies. Public support for environmental protection is limited. Consumerist values still prevail, particularly when environmental protection inflicts serious behavioural changes (Pridham 1994: 81). Some authors have pushed this argument further ultimately embracing the reasoning of the 'Mediterranean Syndrome'. They describe the problem of the southern member states in implementing EU environmental policy as the result of a fundamental 'clash' of political cultures. Southern European countries have political systems traditionally dominated by patronage, clientelism and disrespect for public authority. This Mediterranean culture contradicts the political culture of Northern European countries, which builds on corporate forms of social organization and which also drives EU environmental policies (Yearley, Baker, and Milton 1994: 13; Lewanski 1993; Kousis 1994; Aguilar Fernandez 1994). In a similar vein, Weale et al. conclude from a comparative study of three southern and three northern member states that 'in the southern countries – especially Greece and Spain – there has been a general lack of the kind of civic culture that advances collective interests, and hence promotes environmental culture' (Weale et al. 2000: 245). They observe some recent changes in terms of growing environmental awareness, which, however, has not translated in environmentalism as such embracing behavioural changes in life-style.[3]

Despite its more general approach, the 'Southern Problem' literature suffers from similar explanatory problems as the 'Mediterranean Syndrome' approach. The issue is not so much that the southern countries are diverse in their domestic structures. One could still argue that such differences account for their varying compliance records. Thus, some have claimed that Spain has been more successful in developing mechanisms of horizontal and vertical coordination in environmental policy than the other three southerners. Its administration appears to be more effective and less ridden by corruption and clientelism (Weale et al. 2000: 213-216). Spanish environmental organizations are rather active, and environmental awareness is rising, particularly in some of the regions (Pridham 1994: 67-70; La Spina and Sciortino 1993: 232). If we accepted this assessment, we would expect Spain to have less compliance problems than Italy, Greece, and Portugal, which is confirmed by Spain's lower infringement numbers. Nevertheless, a closer look reveals that there is no direct correlation between certain political, administrative, economic or cultural characteristics of the member states on the one hand, and their non-compliance records, on the other hand.

Differences in institutional fragmentation, administrative style and capacity, financial constraints, and environmental activism do not match the observed varia-tions in compliance performance (table 3). Nor do cross-national variations follow a clear North-South pattern. For example, in Belgium, Germany, Spain, and Italy environmental competencies are strongly decentralized. Whereas vertical frag-mentation is considered a problem for effective implementation in all four coun-tries, non-compliance in Italy and Belgium is significantly worse than in Germany and Spain. Portugal, Greece and Ireland, which are unitary states, may have envi-ronmental ministries too weak to effectively coordinate the implementation of EU policies. But only Greece and Portugal are found among the environmental lag-gards. Italy is the fourth-largest economy in Europe but its level of non-compliance is still higher than in Greece and Portugal, which are the two poorest EU member states. Germany certainly has a stronger administrative capacity than Italy, Greece, Spain, and Portugal. Nevertheless, the German non-compliance record makes the country look more like a laggard than a leader. The reactive and closed administra-tive style of Spain, Greece, Portugal, and Italy may conflict with the more proac-tive, participatory approach of European environmental policies. But the so called 'New Policy Instruments' (NPI), with their emphasis on public participation and transparency, also contradict the German or French administrative tradition of top-down regulation and administrative secrecy and have given rise to serious compli-ance problems in the two countries. Environmental awareness is high with the

---

[3]   To be fair, the authors point out that the distinction between South and North in terms of civic attitudes should not be overdrawn. Even in northern countries, the translation of the support for the environment into behavioural changes can be difficult, particularly if high costs are involved (Weale et al. 2000: 246).

three environmental leaders of Europe: Germany, Denmark and the Netherlands, but only the latter two live up to their reputation when it comes to compliance.

Table 3   Factor constellations and member state non-compliance

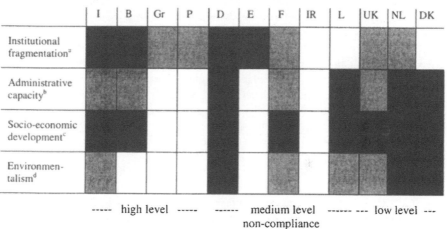

	I	B	Gr	P	D	E	F	IR	L	UK	NL	DK
Institutional fragmentation[a]												
Administrative capacity[b]												
Socio-economic development[c]												
Environmentalism[d]												

----- high level -----   ------ medium level   ------ --- low level ---

non-compliance

The level of non-compliance is measured by the average performance of each member state in the infringement proceedings.

high medium low

a) *Institutional fragmentation* of environmental responsibilities is measured by the degree to which environmental competencies are dispersed between different policy sectors (ministries) and between different levels of government, on the one hand, and the absence or presence of mechanisms of horizontal and vertical coordination, on the other hand (Hanf and Jansen 1998; Weale et al. 2000).

b) *Administrative capacity* to make and enforce environmental policies is measured by government employment as percentage of total employment (personnel), public expenditures for Research and Development (expertise), and dominant policy style (reactive/proactive) (OECD Compendium; Hanf and Jansen 1998; Weale et al. 2000).

c) The level of *socio-economic development* is measured by the average GDP per capita (OECD Compendium).

d) *Environmentalism* is measured in terms of the level of environmental awareness (environment as immediate and urgent problem), the relative importance of the environment as an important problem compared to others, and support for environmental measures requiring behavioural changes, such as taxes on cars and petrol (Eurobarometers EB 51, 1999, B.69; EB 43.1, 1996, 1.1; EB 40, 1994, A. 20-22; EB 46, 1995, 6.1; Weale et al. 2000: Chapter 7).

While the various factors do not systematically co-vary with the different non-compliance records of the member states, it might be a particular constellation of them, which is able to account for the observed cross-country variation. The existence of a peculiar 'Southern Problem' would then require the factor constellations to substantially differ between North and South.

The three (northern) countries with the lowest level of non-compliance do indeed show some similarities in factor constellations (see table 3). Their level of socio-economic development is advanced, they have solid administrative capacities, and the level of environmentalism is significant. Denmark, the Netherlands and the UK mostly differ with regard to their degree of institutional fragmentation. Likewise, Spain, Portugal and Greece share a low level of socio-economic development, administrative capacity and environmentalism while they vary with regard to institutional fragmentation, too. But the three southern member states show different levels of non-compliance. The factor constellations for the two top laggards, Italy and Belgium, are similar but quite different from the three southern countries. In fact, of the four southern countries, Italy's factor constellation fits the southern pattern the least. Its GDP is among the highest of the world. Environmental policy is well institutionalized. Moreover, Italy has a fairly vibrant environmental movement and a well-established Green Party, which formed part of the last centre-left government led by Massimo d'Alemma. Nevertheless, its compliance record is by far the worst among the member states. The factor constellations may suggest a correlation between the level of socio-economic development, administrative capacity, and environmentalism. But they do not fit a clear North-South pattern; nor do they co-vary with the different levels of non-compliance.

In sum, the four southern member states face significant compliance problems. But so do some of the northern countries. Moreover, there is significant variation between the member states, both in the North and the South of Europe, concerning the level of non-compliance and the explanatory factors suggested by the literature. Finally, these differences clearly cut across the North-South divide.

Next to difficulties in explaining cross-country variation, the 'Southern Problem' literature cannot account for sectoral differences either. Member state performance can vary considerably between policies. While most member states appear to have less difficulty in complying with European air pollution control regulations, they run into serious trouble when it comes to other policies, like the Wild Birds, the Fauna-Flora-Habitat or the Urban Wastewater Treatment Directives. Distinct problem pressure, issue salience (Macrory 1992: 365) and differences of efficiency between ministries (Pridham 1996: 71) may partly explain such policy variation. But effective implementation and compliance do not only depend on domestic factors. Several studies point to causes of implementation failure and non-compliance, which are inherent to the structure of European policy-making and the legislation it produces rather than the domestic structures of the member states.

First, EU environmental policies deal with matters of high legal and technical complexity giving rise to questions of interpretation and issues of technical application, as well as difficulties in coordinating the different national authorities in the implementation process. European environmental legislation is based on Directives rather than Regulations. While Regulations are self-executing and automatically enter the legal order of the member states, Directives only set a legal framework and leave it to the member states to enact transposing legislation in the form most appropriate to their domestic context. The regular use of Directives allows for flexibility. But it also creates scope for the member states to delay implementation or to make exceptions. Second, the need for consensus in the Council of Ministers often gives rise to imprecision, ambiguous objectives, and open texture. Poor drafting grants the member states considerable leeway in interpreting and applying European legislation, which may lead to differing understanding and induce the member states to circumvent inconvenient obligations (Collins and Earnshaw 1992: 222-227; Jordan 1999). Third, 'horizontal' policies, such as the Environmental Impact Assessment Directive, cut across conventional administrative boundaries and sectors. Implementation requires the coordination of various public authorities at different levels of government, which contradicts the highly sectorized regulatory structures in many member states (Macrory 1992: 348-349). Fourth, the EU does not possess sufficient monitoring and enforcement capacity (see previous chapter). Finally, European Environmental Law is in want of readily identifiable vested interests willing and able to secure enforcement. The environment is less susceptible to concepts of legal property rights, which induce enforcement by private actors before national courts. Public interest groups often do not have sufficient legal access to commence legal proceedings or prefer to invest their limited resources in mounting political rather than legal pressure on defaulting administrations (Krämer 1989; Macrory 1992).

While such policy factors are important, they alone do not explain implementation failure and non-compliance either. If compliance problems essentially arose from the specific nature of EU environmental policies, all member states should exhibit similar implementation deficiencies, which is clearly not the case. Some scholars seek to integrate policy and country variables in explaining the differing levels of member state compliance with EU Environmental Law (Haas 1998; Mendrinou 1996). Yet, such 'integrative' approaches hardly specify how European and domestic factors interact. Nor can they explain variations in non-compliance across different policies within one single country. Why does a member state fail to comply with some EU environmental laws while it is more successful with others? What is needed is an approach that not only allows to explain variation in compliance *across* the different EU member states, especially across the North-South divide, but also accounts for variation in compliance with different policies *within* member states. The next section develops such a model by systematically linking European and domestic factors in explaining compliance with EU environmental policies.

Pressure from Below and from Above: The Pull-and-Push Model

The Pull-and-Push Model is based on two major propositions. First, compliance problems only arise if the implementation of European policies imposes considerable costs for the member states. The less a European policy fits the regulatory structure of a member state, the higher the adaptational costs in implementation and the lower the willingness of public and private actors to comply. Second, the willingness and/or ability of public and private actors to bear the costs of implementing poorly fitting EU policies is influenced by pressure for adaptation from 'below', where domestic actors mobilize against ineffective implementation (pull), and from 'above', where the European Commission opens infringement proceedings (push).

The Pull-and-Push Model is based on a rationalist account of actors' behaviour. Compliance is conceived as a rational decision where actors weigh the costs against the benefits of compliance. Positive incentives and negative sanctions are crucial in inducing compliance with 'misfitting' policies since they are able to influence the cost-benefit calculations of actors. Such an approach ignores socio-logical institutionalist explanations of rule-consistent behaviour, which conceptu-alize compliance as the socialization of actors into new norms and rules through processes of arguing, persuasion and social learning as a result of which actors redefine their interests and identities. Actors comply because they believe it is the appropriate thing to do, irrespective of the costs and benefits involved (cf. Börzel and Risse 2002; Checkel 2001a). This is not say that socialization is irrelevant to inducing compliance. The Pull-and-Push Model identifies *one* causal mechanism in explaining compliance. It specifies conditions under which non-compliance with EU environmental policies is likely to occur. While policy misfit is a general (nec-essary) condition for non-compliance, pressure from below and from above is only one possibility to overcome the resistance of actors to face the costs.

Unlike most of the literature, the Pull-and-Push Model focuses on compliance rather than non-compliance. It takes policy misfit as the necessary conditions for compliance problems. However, pressure from below (pull) and from above (push) explains why 'misfitting' policies do not necessarily result in compliance failure. Choosing compliance as the dependent variable helps to avoid the methodological problem of using the same measurement for the dependent and the independent variable (endogeneity). Infringement proceedings are essential for generating pres-sure from above (push). At the same time, they serve as one indicator for non-compliance. The Pull-and-Push Model not only tries to work around the endogene-ity problem by explaining compliance rather than non-compliance. It also develops compliance indicators that are not related to infringement proceedings (see chapter 4).

Policy Misfit as the Necessary but not Sufficient Cause of Non-Compliance

Policy Misfit is the result of incompatibilities between European and national poli-cies. The more EU policies challenge or contradict corresponding policies at the national level, the higher the misfit and the greater the need for a member state to adapt its legal and administrative structures in the implementation process (Duina 1997; Knill 1998). Such legal and administrative changes may incur high costs, both material and immaterial, which public and private actors are hardly inclined to bear. The higher the costs of adaptation (misfit), the more likely are compliance problems to occur due to the resistance of domestic actors to deal with the costs. In case of fit, by contrast, the EU policy can be easily absorbed into the existing legal and administrative structure and does not give rise to any compliance problems.

Policy misfit is defined as the incompatibility between the problem solving approach, policy instruments and/or policy standards of an EU policy and the corresponding policy at the national level. Only if an EU environmental policy challenges one (two or all) of these three elements, its implementation gives rise to problems for a member state because of the adaptational costs incurred.

Problem solving approach refers to the general understanding of an administration of how to tackle problems of environmental pollution. Two ideal types of problem solving approaches can be conceptualised:[4]

• *a precautionary, technology- and emission based approach*, which prioritizes the prevention of harmful effects for the environment by imposing legally binding standards to be uniformly applied by all polluters irrespective of the differing local quality of the environment and the application of the available technology irrespective of the cost involved compared to the potential benefit for the environment;

• *a reactive, cost/benefit- and quality based approach*, which aims at reducing harmful effects for the environment by the setting of quality standards for a certain area, which allows for a more flexible application to be negotiated with the regulated parties where the costs of a technology are balanced against potential environmental improvements.

Policy instruments refer to the techniques applied to reach a policy goal by inducing certain behaviour in actors. They can be classified according to the following dimensions:

• *regulatory, command and control instruments*, which regulate behaviour through prescriptions and prohibitions threatening negative sanctions in case of non-compliance vs. *market-oriented instruments* offering financial incentives and *participatory, communicative instruments* providing information and encouraging public participation and deliberation;

• *substantial regulation* by legally binding standards vs. *procedural regulation* through procedures, such as the balancing of costs and benefits or the public participation in authorization procedures.

Policy standards, which can be quantitative or qualitative in nature (see above), refer to the guiding values set by a policy, e.g. for air or water quality.

Costs, which arise in the adaptation of domestic regulatory structures to 'misfitting' European policies, can be of differing nature:

---

4   The classification draws on the categories developed in a research project on the impact of national administrative traditions on the implementation of EU environmental policy conducted at European University Institute, Florence, in April 1997 (Knill 1997; cf. Héritier, Knill, and Mingers 1996).

- *economic costs* referring to the investment of new, and the reallocation of existing resources, such as the acquisition of additional personnel, expertise, monitoring and abatement technology;
- *political costs* concerning the support of electoral constituencies (endangering jobs) or political credibility and reputation (being a 'good European' or an 'ecological leader');
- *cognitive costs* related to challenges of certain values or beliefs ingrained in problem-solving approaches and standard operating procedures of public administrations (administrative confidentiality vs. access to information and transparency; end-of-the-pipe vs. polluter-pay).

Yet, policy misfit causing costs of adaptation does not necessarily lead to implementation failure and non-compliance. The mobilization of domestic actors, who pull the policy down to the domestic level by pressuring the public administration to properly apply it, may persuade national public actors 'to give priority to environmental policy and to embrace new directions' (Pridham 1994: 84). Legal action or public campaigns, in which environmental groups denounce a member state administration for not complying with EU legislation, provide an additional incentive for better compliance. Such domestic mobilization often triggers external pressure for adaptation from 'above' by the European Commission, which opens infringement proceedings against recalcitrant member states.

Domestic Pressure for Adaptation from 'Below': The Pull-Factor

Domestic pressure for adaptation arises from domestic mobilization. EU policies usually have direct effects on domestic actors, imposing constraints for some and offering opportunities to others. It is characteristic for (EU) regulatory policies that, unlike distributive policies, they inflict costs in the implementation rather than in the formulation and decision-making stage of the policy process (Majone 1993). Domestic policy-makers, administrators, and regulated parties, who have to bear the costs of EU environmental policies, tend to resist their implementation. This resistance, however, can be counterbalanced by other domestic actors who strive to pull down the EU policy to the domestic level by pressuring the public administration to legally incorporate, practically apply and enforce it. Moreover, domestic actors may exert direct pressure on regulated parties to comply with European regulations. Domestic pressure on public and private actors can materialize through various channels.

First, political parties have considerable scope for mobilizing public opinion and influencing public attitudes. They can raise concerns about the proper implementation of policies vis-à-vis the government. When in Germany the Green Party became the junior partner in a coalition government with the social democrats in 1998, environmental interests not only got better access to the policy arena. With a green environmental minister, it has become more difficult for the German government to justify the continuous non-compliance with certain European environmental policies, such as the Fauna-Flora-Habitat Directive of 1992. One of the first environmental actions of the new red-green coalition was to bring the Nature Protection Act on its way, which the previous government had driven into political stalemate.

Second, societal actors can act as 'watchdogs' drawing the attention of both public authorities (national and European) and the public opinion on incidents of non-compliance with EU environmental legislation. Since the influence of national environmental organizations on European policy-making tends to be limited, they concentrate on the legal implementation of EU Directives in their country lobbying national policy-makers for a correct and complete transposition into national law. When it comes to practical application and enforcement, citizens and environmental groups are an important source of information for the European Commission, which lacks proper monitoring capacities. Moreover, societal actors, in alliance with the media, often launch public campaigns to shame their government for serious instances of non-compliance. Finally, societal actors play a major role in the decentralized enforcement of European Law through national courts. Given its supremacy and direct effect, European legislation confers rights to any affected individual to challenge non-compliance before national courts. Since the early 1900s, individuals can even claim financial compensation for damages which their government's failure to effectively implement European Law inflicted on them (Craig 1993, 1997).

Third, the media can play an important role for domestic mobilization. Media coverage often decides whether an environmental issue gains public attention and support. It is often crucial for the success of environmental campaigns launched by citizen groups and public interest organizations. High media salience of an issue may also trigger a public debate on political responsibilities for non-compliance and increase the pressure for political as well as private action to rectify the situation. Reports on environmental issues, finally, can further environmental awareness, particularly when they are not exclusively geared towards emergency situations and scandals.

Finally, economic actors can mobilize in favour of compliance with a policy. Being the target of most environmental regulations, business and industry tend to be hostile to rather than supportive of environmental protection. Yet, green industry and multinational corporations have realized that common standards at the European level may be preferable to lower average standards that vary from one country to another and are subject to unpredictable changes. Moreover, companies operating in countries with higher environmental standards may benefit from rising general standards because they are already in compliance. Likewise, lax implementation and enforcement in other countries can cause them significant competitive disadvantages. Thus, companies have made extensive use of the Article 234 preliminary ruling procedure litigating against what they perceive as an unfair application of European Law by national authorities, which tends to favour their competitors (Cichowski 1998).

Domestic mobilization is most effective if it is able to link-up with the European Commission, which can exert additional pressure for adaptation from above by initiating infringement proceedings.

External Pressure for Adaptation from 'Above': The Push-Factor

The EU has only limited enforcement powers. Nevertheless, the infringement proceedings of Article 226 as described in the previous chapter provide the European Commission with a means of exercising external pressure on non-compliant member states by increasing their costs of non-compliance.

Infringement proceedings can incur significant political costs on the member states. The highly regulated countries in particular do not wish to damage their reputation as 'environmental leaders'. Being prosecuted for non-compliance with EU environmental standards could not only irritate their domestic constituencies. It might also undermine the political credibility of environmental leaders in pushing for new or higher environmental standards at the European level when they do not comply with existing regulations. But environmentally less advanced countries are concerned about their reputation, too. Being perceived as a cheater and free-rider undermines the bargaining power of leaders and laggards alike. Moreover, member state government may wish to demonstrate that they conform to the group of states to which they belong and whose esteem they care about (Finnemore and Sikkink 1998). Violating EU law can seriously contradict the image of being a 'good European', which some countries seek to convey.

While the Commission prefers bilateral negotiations over the threat of sanctions, member states' fears of being publicly named and shamed for non-compliance with European Law gives the Commission a powerful 'stick' to discourage non-compliance. The initial stages of the infringement proceedings are informal and kept strictly confidential. With the exception of non-transposition (see previous chapter), Formal Letters are only published if the Commission wishes to exert specific pressure on a member state. Whereas Reasoned Opinions are released, they still get little public attention. But national media becomes increasingly interested when cases are referred to the European Court of Justice. ECJ rulings are usually reported, at least in the print media. Cases of second referral for non-compliance with ECJ rulings, finally, receive broad coverage, particularly since they may result in financial penalties, like in case of Germany's non-compliance with the Environmental Impact Assessment Directive in December 2000, where the Commission asked the ECJ to impose a daily fine of 237,000 Euro. The Commission considers the threat of financial sanctions as a success so far (Commission of the European Communities 2000: 11). Of the eight cases where it had asked the ECJ to impose financial sanctions all but three got settled before a court ruling. The combined effect of material and reputation costs of recalcitrant non-compliance appears to have a strong dissuasive effect on the member states. This is also true for the infringement proceedings in general. External pressure of the Commission, as it increases throughout the different stages of the infringement proceedings, proves rather effective in discouraging persistent non-compliance with European Law (cf. Jönsson and Tallberg 1998).

The Combined Pressure from Below and from Above

Pull and push factors may work independently in counteracting non-compliance with 'misfitting' EU policies. But often they interact and, indeed, are most effective if they appear combined. Many infringement proceedings are triggered by citizens, public interest groups, or companies who lodge complaints with the European Commission. While domestic actors fulfil important monitoring functions for the Commission, European institutions offer them an authoritative venue to challenge non-compliant state behaviour. Despite its limited monitoring capacities, the Commission may be more responsive to domestic concerns about non-compliance than national institutions. Administrative and legal proceedings are costly and time-consuming. Moreover, national courts are not necessarily supportive of the litigation by domestic actors, as the jurisdiction of German administrative courts on the application of the Environmental Impact Assessment Directive shows (see chapter 4).

Domestic actors are not only empowered by the possibility to trigger external pressure for adaptation. Their domestic lobbying, shaming and litigation activities complement and reinforce pressure from above. Often, environmental organizations pursue a twofold strategy of lodging a complaint with the Commission and of taking political and legal action at the domestic level. At the same time, the political costs of European infringement proceedings are significantly higher for a member state when domestic actors manage to trigger a public debate on the issue and domestically shame their government. In sum, if public authorities get 'sandwiched' between adaptational pressure from below and above, EU environmental policies that incur high adaptational costs are more likely to be effectively implemented and complied with. The 'pull and push' effects are summarized in figure 3.1.

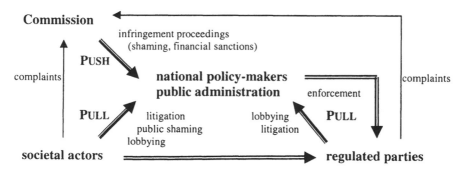

Figure 3.1 The Pull-and-Push Model

The Pull-and-Push Model generates the following hypothesis about when 'misfitting' policies are likely to be effectively implemented and complied with despite the adaptational costs they incur:

> *Compliance with policies that incur significant adaptational cost (policy misfit), is the more likely, the higher the adaptational pressure from below and from above (domestic mobilization, infringement proceedings).*

Explaining variations in compliance as the result of differing degrees of policy misfit and of pressure for adaptation systematically links policy and structural variables at the European and the domestic level.

Policy misfit is not merely a function of some policy variables at the European and domestic level, which prove to be incompatible. It is influenced by the structure and dynamics of European policy-making as well as by certain structural features of the member states. EU environmental policy-making can be described as a regulatory contest among the member states, which compete for EU environmental rules and regulations that conform to their own interests and regulatory principles and policies (see below). Member state diversity with respect to regulatory preferences and action capacities to pursue these preferences as well as the need for consensus and coalition-building in EU decision-making prevent a situation where one member state single-handedly manages to shape EU environmental policies according to its regulatory style and tradition. In the absence of a consistent regulatory framework, each member state, high- or low-regulating, can face significant policy misfit. The next section, however, will show that environmental late-comers in the South are policy takers and therefore face higher adaptational pressure than environmental first-comers in the North, which are the makers of EU environmental policies.

Pressure for adaptation from below is mainly a function of structural variables at the domestic level. The capacity of domestic actors to generate internal pressure for adaptation largely depends on their organizational, political and legal resources (Kitschelt 1986; Kriesi 1991). The institutions of the member states offer different opportunities and constraints to their domestic actors, which they can explore to mobilize in favour, or against, the effective implementation of European environmental policies. Whether domestic actors mobilize, however, is not merely a question of their action capacities. The decision on how to invest their scarce resources also depends on the problem perceptions and political priorities of domestic actors. Both are influenced by the level of socio-economic development which a country has reached. Heavy industrialization not only causes different problems of environmental pollution. Highly industrialized societies tend to be more supportive of environmental protection measures than societies, which are still concerned with generating economic growth rather than fighting its environmental consequences. Pressure for adaptation from below may therefore vary across policies within one country, depending on whether an EU policy helps to address environmental problems that domestic actors perceive as pressing. The next section will argue that domestic pressure is lower in southern member states not only because their political opportunity structures are less open than in the North but also because the regulatory contest in EU environmental policy-making tends to produce policies that are mostly oriented towards the environmental problems and economic concerns of the northern industrial first-comers.

Reformulating the 'Southern Problem': High Policy Misfit and Low Domestic Mobilization

By integrating European and domestic factors to explain compliance, the Pull-and-Push Model cannot only account for cross-country and cross-policy variation. It also allows to reformulate the 'Southern Problem' in more general terms by identifying high policy misfit and low domestic pressure for adaptation as the major causes of lower compliance in the South. In a nutshell, southern countries face greater compliance problems because the 'leader-laggard' dynamics of EU environmental policy-making tend to produce higher policy misfit for them while their closed political opportunity structures and lower level of socio-economic development discourage the domestic mobilization necessary to make the pull-and-push mechanism work.

High Misfit: The 'Leader-Laggard' Dynamics in EU Environmental Policy-Making

European environmental policy-making has been characterized by a 'leader-laggard' dynamics where the leaders push the Community process along drawing the laggards up to their levels of environmental protection (Haas 1993; Héritier 1994). The literature views this 'push-pull' process (Sbragia 2000) as a major factor in promoting environmental progress at the European level. What is often overlooked, however, is that the 'leader-laggard' dynamics entail a strong element of regulatory contest which produces EU environmental policies that undermine both the capacity and the willingness of southern member states to comply with EU environmental rules and regulations.

Northern leaders Regulatory contest in EU environmental policy-making is driven by a general incentive for the member states to upload their national policy arrangements to the European level in order to avoid adaptational costs in the implementation process, both for public administration and business (Héritier, Knill and Mingers 1996).[5] Due to their distinct social, political, and economic institutions, the member states substantially differ, first, in their policy preferences, and second, in their capacity to pursue these preferences at the European level. Both policy preferences and action capacities are closely linked to the level of economic development, particularly in the area of regulatory policies. Highly industrialized and urbanized states in Northern Europe have developed strict, highly differentiated legal regulations of environmental pollution in all media accompanied by just as highly differentiated state implementation arrangements. They have a strong incentive to harmonize their demanding standards at the European level. First,

---

[5]   Adrienne Héritier uses the term 'regulatory competition', which, however, may be misleading since it is commonly associated with an assumption in economic theory that market liberalization leads to increasing competition between systems of different economic and political institutions.

high-regulating states seek to coordinate their efforts in fighting environmental pollution where its transboundary nature renders their unilateral measures less effective. European air pollution control, for instance, started as the attempt of some member states to fight massive forest dying caused by the long-range transport of certain air emissions, such as sulphur dioxides (SO_2) and nitrogen oxides (NO_x).

Second, high-regulating states wish to obtain favourable competitive conditions for their domestic industry and avoid environmental dumping in low-regulating member states. Thus, German companies have repeatedly complained that they must invest more into environmental measures than any of their European competitors. Environmental measures often raise the costs of production and may result in competitive disadvantages with comparable sectors in other countries. Chemical firms, for instance, have pressed the German government to spread their higher costs of compliance with environmental standards by means of stricter European laws across the other member states. Likewise, German car manufacturers strongly opposed the setting of speed limits to fight air pollution and supported the mandatory imposition of catalytic converters throughout Europe, in which they had invested in order to export to the American market (Aguilar Fernández 1997: 105). Pan-European multinational firms often support the harmonizing of strict standards since it is in their operational interest to have only one set of EU rules to comply with rather than 15 different sets of national regulations.

Third, high-regulating states hope to reduce adaptational costs in the implementation of European environmental policies. Incorporating 'alien' elements into a dense, historically grown regulatory structure that is ingrained in a particular state tradition can impose considerable costs, both material and cognitive. For instance, procedural regulations, like the Integrated Pollution Prevention and Control Directive, that prescribe the cross-media evaluation of potentially harmful activities contradict highly sectorized administrative structures with their medium-specific approach as they have historically developed in many countries. The incorporation of such integrated policy measures requires comprehensive legal and administrative adaptations to preserve the consistency of the regulatory framework. Adaptation is often also difficult because administrators tend to be reluctant to give up traditional problem-solving approaches and policy instruments, which they consider as proven effective in fighting environmental pollution (Knill and Lenschow 2000a).

Fourth, national governments are anxious to respond to the 'green' demands articulated within their political systems. In face of substantial environmental degradation and dramatic accidents, such as Seveso and Chernobyl, environmental awareness and societal activism is high in industrial societies. They expect their governments to push for effective environmental regulations at the European and the international level. Finally, high-regulating states may have an interest in expanding technology markets for their own industry. Compliance with strict environmental standards provides industry with a powerful incentive to develop and improve 'green' technologies, which can then be exported as 'best available technology' to lower-regulating countries to help them adapt to more stringent European standards (cf. Héritier, Knill and Mingers 1996: 10-15, 23-28).

Setting the pace of EU environmental policy-making not only presupposes established domestic policies but also the capacity to push them through the European negotiation process, often against the opposition of other member states with diverging policy preferences. This is not merely a question of voting power in the Council of Ministers, particularly since qualified majority voting has become more prevalent. Smaller countries have often effectively shaped European policies. Denmark, for instance, managed to transform its national Plan for the Aquatic Environment into the Urban Waste Water and Nitrate Directives (Andersen and Liefferink 1997: 14), while the Netherlands convinced the other member states to adopt its high standards for small car and truck emissions (Axelrod and Vig 1998: 77). Offering expertise and information to the European Commission in the drafting of policy proposals is a very effective way of injecting national preferences into the European policy process (Zito 2000). Another is the strategic employment of national environmental bureaucrats in Brussels for up to three years. The Commission also asks the member states to second experts with specific knowledge to help prepare Directives (cf. Héritier 1994; Liefferink and Andersen 1998b: 264-266). Coalition-building and interest accommodation skills provide a third important source of influence in shaping European policies. The Dutch have a particular reputation for working out complicated, tailor-made compromises, which are acceptable to all parties involved (Andersen and Liefferink 1997: 27-28). Being present in the various networks that prepare and accompany the European negotiation process demands considerable manpower, expertise and information, which the member states have not equally available. The German Ministry of Environment counts some 900 employees and the Federal Environmental Office has another 850 experts working in the field. The Dutch core environmental administration employs about 1500 persons (Hanf and Gronden 1998: 171).[6] Not surprisingly, German and Dutch experts are 'omnipresent' in the environmental committees and hearings of the Commission, the Council, the European Parliament, the Economic and Social Committee or the Committee of the Regions. High-regulating countries cannot only offer practical and technical expertise and information to European policy-makers. The Commission as the agenda-setter of EU policies is rather responsive to policy demands of environmental forerunners since their stringent legislation could distort trade, thereby jeopardizing the functioning of the Single Market (Sbragia 2000: 240). Thus, the EU, which had already adopted a Directive on beverage containers in 1985, passed the Packaging and Packaging Waste Directive in 1994 to accommodate the recycling of other materials. The Directive was a reaction to emerging Danish, Dutch, and German legislation that could develop into potential trade restrictions (Haverland 1999). Likewise, Germany successfully pushed for stringent car emission exhaust regulations at the European level by

---

[6]   To put this into perspective, the Italian Environmental Ministry, for instance, employs about 400 people (Lewanski 1998: 139) and the French 500 (Larrue and Chabason 1998: 68).

threatening to introduce US standards unilaterally (Andersen and Liefferink 1997: 15).

While they share a common interest in harmonizing their environmental standards at the European level, high-regulating countries significantly differ with respect to their regulatory structures as a result of which they often compete for European environmental rules and regulations that conform to their own interests and regulatory principles (Héritier 1996). Britain and Germany, for instance, adhere to opposite problem-solving approaches in how to best protect the environment. The German precautionary approach emphasizes strict emission standards that can be only complied with by applying the best available technology. The British approach, by contrast, is more reactive. It relies on quality standards and allows for a weighing of the economic costs against the ecological benefits of a policy. The two problem-solving approaches, which imply the use of different policy instruments, are hard to reconcile. As a result, even high-regulating member states can face significant policy misfit, if a member state with a competing regulatory approach carries the day in the EU policy process. The European Large Combustion Plant Directive of 1988, for instance, follows the German approach and thus imposed significant costs on the UK (Boehmer-Christiansen and Skea 1991). British administration was forced to adapt its regulatory approach, which has traditionally been based on voluntary regulation and negotiation with industry. It had to give way to more formal regulation. While German industry gained a competitive advantage because it already applied best available technologies to reduce harmful emissions, the British had to invest in new abatement technologies whose major producers happened to be German companies (cf. chapter 4).

Since EU decision-making entails a need for consensus and coalition building, no single member state is able to ultimately win the regulatory contest and systematically shape EU environmental policies according to its regulatory style and tradition. Even if a member state manages to frame a policy proposal according to its domestic arrangements, it cannot entirely control the dynamics of the European negotiation process. Germany, for instance, successfully up-loaded its regulatory approach for drinking water. The European Drinking Water Directive of 1980 was largely modelled after the German regulation. Yet, some of the European standards became more stringent than the German ones as a result of which it took Germany more than 12 years and a conviction by the European Court of Justice to comply with the Directive (cf. chapter 4). Major proposals for environmental legislation are subject to substantial intergovernmental bargaining in the Council of Ministers. European policy decisions not only require the consent of a certain number, if not of all member states (e.g. eco-tax) but have to accommodate the position of the European Parliament, whose co-decision powers have become considerably strengthened over the last years and which usually takes a 'green' position. Given the dynamics of the regulatory contest, European environmental legislation does not provide a coherent regulatory framework but looks like a 'patchwork' (Héritier 1996), which produces significant policy misfit for all member states and thus explains cross-policy variations within member states. Yet, since the southern latecomers are policy takers rather than policy makers, which lack both the policies and the capacity for uploading, they are much more likely to face policy misfit than the northern first-comers and forerunners.

Southern laggards The Southern European member states are industrial and environmental late-comers as a result of which they have distinct policy preferences that substantially diverge from their northern counterparts. Unlike the latter, their overriding priority has been economic development and growth to the extent of subordinating the environment, although this is somewhat less true for Italy than for Spain, Portugal and Greece. They wish to keep the level of environmental regulation low, and consequently, tend to oppose stringent environmental standards at the European level (Weale et al. 2000: 95-97).[7] First, less exigent regulations constitute a comparative advantage over high-regulating countries due to lower production costs while complying with stringent protection requirements places a heavy financial burden on industry and hampers its efforts to catch up with industrialized countries. Second, strict European standards do not provide new sales opportunities. On the contrary, they may work as trade barriers until low-regulating countries have adapted. Moreover, in order to achieve compliance, companies often have to import environmental technology from environmental first-comers. Third, late-comers are particularly averse to the costs of environmental policies since increasing the standard of living and generating employment are still the overriding priorities on their political agenda. Environmental protection is often viewed as having a sharp trade-off against economic development and employment creation. Fourth, since environmental awareness and ecological activism is only emerging in these countries, policy makers feel less pressure to take environmental measures and gain little support for enforcing costly policies. While public concern for the environment is rising, people are still reluctant to accept protection measures that may impose significant costs. Finally, building-up regulatory structures is often even more expensive than fitting European policies into historically grown, comprehensive domestic arrangements. New administrative units, procedures, and technologies have to be established for the practical application and enforcement of European policies. Additional investments are often necessary to build-up staff-power with the necessary technical and scientific knowledge to apply environmental regulations and assess compliance at the local level.

---

[7] In their analysis of member state preferences towards more stringent environmental standards at the EU level, Weale et al. argue that the division between northern and southern countries is less clear-cut than suggested by the leader-laggard dynamics. While there is a clear leader group (Denmark, the Netherlands, and Germany), they cannot identify a coherent core group of laggard countries. Greece, Italy, the UK, and France tend to oscillate between the two groups depending on the issue involved. But they also concede that the varying national preferences are strongly influenced by the comparative economic standing of the member states and the extent to which they felt capable of 'securing the resources to finance the investment that the standard would require' (Weale et al. 2000: 97).

Besides diverging preferences, the southern late-comers face a double disadvantage in the European regulatory contest vis-à-vis the northern first-comers. Since their regulatory structures are less developed, they hardly have any policies to up-load to the European level. But even if they had the policies, they would lack sufficient staff-power, money, and expertise to actively shape European policies. Late-comers only have a limited number of administrators and scientific experts, and they cannot draw on a long-time experience of fighting environmental pollution (Font and Morata 1998; Spanou 1998). Thus, they have little knowledge and information which they could contribute to the development of European solutions to transboundary problems, such as air pollution or toxic waste – problems which they have only started to experience with their industrialization taking full speed. Consequently, their national networks in the European policy-making bodies, consisting of both permanent and temporary staff, are small compared to those maintained by the environmental pace-setters. Instead of shaping environmental policies, environmental late-comers tend to emulate policy solutions from highly-regulating countries. Before Greece, Portugal, and Spain joined the European Community in the first half of the 1980s, their environmental policies had been only weakly developed. European policies became the determining factor of their environmental regulatory structures through the down-loading of EU Directives (Pridham 1996; Font and Morata 1998; Spanou 1998).

Since late-comers have neither an incentive nor the capacity to push or support strict European measures, they try to block or delay policy initiatives hoping to reach at least some temporary exemptions (derogations), financial compensations (side-payments) or concessions in other issue areas (package deals). In the negotiations on the Single European Act, the southern member states and Ireland linked their support for the Single Market, with its implied higher standard of environmental protection, to the issue of the structural funds (Weale et al. 2000: 45). Likewise, Spain, together with Ireland, rejected the extension for qualified majority voting on environmental measures, pushed for by the northern pace-setters in the negotiations on the Maastricht Treaty, claiming that European policies had shown no concern for the problems of economically less developed regions (Aguilar Fernández 1997: 108). To balance EU environmental policies which they perceived as biased towards the worries of the industrially more advanced member states, Spain and the other late-comers insisted on provisions of financial support for member states which could not easily bear the costs of environmental protection. Particularly Germany strongly opposed the idea of an environmental fund. The conflict was ultimately resolved in a package-deal. The late-comers' approval of extended qualified majority voting was literally bought by the establishment of the Cohesion Fund of which about 50 per cent should be dedicated to financing the implementation of EU environmental policies in Spain, Portugal, Greece, and Ireland.

Financial assistance granted by the Cohesion Fund and other EU environmental programmes might cover up to 75 per cent of the compliance costs. But this is often not sufficient to compensate for the weak implementation capacities of environmental late-comers. The Urban Waste Water Treatment Directive of 1994, which was a German initiative, is a case in point. It obliges the member states to provide urban agglomerations with systems collecting and treating urban waste waters. Adequate facilities were largely missing in the low-regulating countries. In Spain, two-thirds of the treatment facilities did not comply with the requirements of the Directive. To finance the building of new facilities and the upgrading of existing ones, the Spanish Water Treatment Plan of 1995 envisioned public investments of about EUR 10.8 billion, which is five times more than Spain had invested in waste water treatment between 1985 and 1993. The Cohesion Fund will only cover parts of the costs.

The regulatory contest in European policy-making exacerbates the capacity problems of industrial and environmental late-comers. They wish to catch-up with industrialization in other member states. But at the same time, they have to download European policies and build-up regulatory structures for fighting environmental pollution that most of the high-regulating countries only established after they had completed their industrialization process. High costs and low capacity make it often difficult, if not impossible, for late-comers to comply with European Environmental Law.

But the 'leader-laggard' dynamics have not only caused serious capacity problems for the South in complying with EU environmental policies. It also challenges the political legitimacy of EU Law and, hence, the political willingness of southern member states to comply. The southern late-comers tend to view EU environmental legislation as imposing the standards of the 'rich north' on the 'poor south' (Aguilar Fernandez 1993: 232-233; Yearley, Baker, and Milton 1994). From their very inception, EU environmental policies have exhibited environmental concerns of the northern industrialized countries such as air pollution, waste disposal, control of chemical substances, or nuclear energy. These policies do little to resolve some of the most pressing environmental problems of the South, which concern aridity, water shortages, soil erosion, land degradation, and forest fires (Aguilar Fernandez 1994; cf. Baldock and Long 1987). The European Community, for instance, enacted strict regulations for water quality. But southern member states suffer more than anything else from problems of water quantity. Investing in expensive measurement technologies for monitoring water quality may deprive these countries of resources they could use for fighting derogation and desertification. If environmental late-comers increasingly feel compelled to implement policies that are costly, fail to account for their economic concerns and do not help solving their most pressing environmental problems either, the regulatory contest in EU policy-making not only weakens their capacity but also their willingness to comply with stringent environmental regulations.

Low Domestic Pressure: The 'Pull-and Push' Dynamics in Domestic Environmental Policy-Making

Due to the 'leader-laggard' dynamics in EU environmental policy-making, southern late-comers are more likely to face significant policy misfit than the northern forerunners. The result is a somewhat paradoxical situation where those member states with the most limited policy-making capacities have to bear the highest compliance costs. If the South has not sufficient resources to actively shape EU policies, the lack of money, staff-power, expertise, and administrative coordination is even more manifest in the implementation process. But not only do the southern member states have fewer resources to cope with higher compliance costs than their northern counterparts. Their governments and administrations also feel less pressure to redirect existing resources. Northern governments are equally averse to costs imposed by 'misfitting' EU policies. But they are more likely to face pressure from domestic actors, which may counteract attempts of cost avoidance and cost evasion. In the South, by contrast, closed political opportunity structures combined with a lower level of socio-economic development seriously constrain the capacity of societal actors to pull down EU policies.

The impact of societal actors on policy outcomes largely depends on the political and legal resources provided to them by the political opportunity structures within which they operate (Kitschelt 1986; Kriesi et al. 1995; Tarrow 1998). The formal and informal institutions of a political system offer domestic actors access points or venues (Baumgartner and Jones 1991) through which they can seek to influence the policy-making process.[8] Despite cross-national variations, the political systems of northern member states are in general more open and accessible to environmental interests than those of southern countries, where traditional statism and the weak institutionalization of environmental politics bread rather closed political opportunity structures (Aguilar Fernandez 1993; cf. Weale et al. 2000: chapter 7). Government authorities dominate both decision-making and implementation processes. Institutionalized relationships of cooperation and consultation between public actors and representatives of economic and societal interests are weak. Unlike in most northern countries, the regular consultation and participation of experts is not a customary practice either. A strong legacy of state interventionism and authoritarianism induces public authorities to adopt an exclusive rather than integrative strategy towards environmental lobbying activities, which impairs informal access to the policy process (Kriesi 1991: 222-223).

The weak institutionalization of environmental policy further restricts the influence of environmental interests on policy outcomes. It also reflects the still limited extent to which environmental concerns are taken seriously by the Southern European member states (Weale et al. 2000: chapter 6). They have begun to establish state secretaries or ministries of environment, environmental protection agencies, advisory boards, and communal public services. But the power of environmental authorities is still weak. They have to share environmental responsibilities with a variety of other ministries and departments (e.g. public works, energy, industry),

---

[8]   Violent and illegal forms of social mobilization, such as acts of civil disobedience, occupations, blockades, riots, or physical attacks against persons, are left aside here since the theoretical approach of this study focuses on ways in which domestic actors make (joint) use of institutional opportunities at the EU and the national level to pressure public authorities to effectively implement costly EU policies.

which are often more resourceful and represent powerful clienteles. The share of state expenses for measures to improve the quality of the environment remains low.

Closed political opportunity structures combine with a lower level of socio-economic development and result in weak public support for environmental issues. Economists have argued that environmental quality is a superior good, for which demand rises as income grows (cf. Hirsch 1977: chapter 3). In a similar vein, political sociologists have observed that in prosperous countries post-material values or new political issues, such as environmental protection, have greater resonance among the public because its elementary needs are largely satisfied and people become less interested in quantitative improvements of their material standard of living (Inglehart 1990). In late-industrializing countries, by contrast, material values still prevail. In short: the richer people are, the more environmental quality they want. Environmental interests have little political impact if the general public, or at least a significant part of it, does not support their concerns, or worse, has opposed preferences (Burstein 1999). The link between economic development and environmental awareness has been confirmed by opinion polls, which find that the concern for the environment is significantly lower in southern countries as compared to their northern counterparts (Rucht 1999: 217-218; Weale et al. 2000: 237-246). Recent data show a more mixed picture (see chart 3). Greece, Portugal and Italy display the highest level of absolute concern for the environment. The majority of the Southern Europeans seem to be willing to support even costly measures to fight environmental pollution, such as higher taxation on private vehicles and fuels (Eurobarometer EB 46, 1996, B.66-67). However, when asked about the relative importance of the environment compared to a whole range of other problems, environmental awareness is still relatively low in the four southern countries. So far, public concern for environmental degradation in the South has not been strong enough to become the focus of a specialized party and to render this party electorally relevant so that established parties would feel the need to take this concern systematically into account. With the exception of Italy, which has a proper ecologist party (*I Verdi*),[9] main parties of Southern European countries have only paid attention to environmental issues on an *ad hoc* basis (Weale et al. 2000: 246-256).

---

[9]   But note that the Italian Greens owe much of their electoral success to a general frustration with the Italian party system.

**Chart 3 Absolute and relative significance of environment
as an important problem for the EC 12**

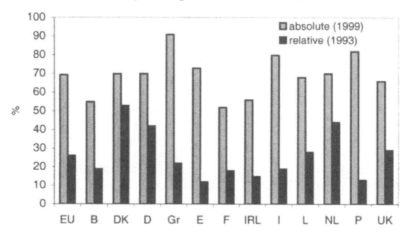

Source: Own elaboration after Eurobarometers 51 (1999, B.69) and EB 40 (1994, A.22).

A lower public concern for the environment also limits the possibilities of environmental interests to exert pressure through media action (press conferences, public resolutions) and public campaigns (petitions, manifestations, referenda). Media coverage of environmental matters is irregular and often reactive to emergency situations and scandals. Moreover, environmental problems, which enjoy high issue salience, are not necessarily those which are addressed by EU environmental policies (see above). Diverging problem pressures and problem perceptions may seriously constrain the capacities of environmental actors to mobilize the public against violations of European Law. This is all the more true since European environmental protection measures are often perceived as conflicting with concerns of the public and vested economic interests about creating and securing employment and economic growth. Complying with European pollution control standards, for instance, requires the southern industry to heavily invest in abatement technologies. Such environmental investments significantly increase production costs undermining a major competitive advantage of Southern European companies vis-à-vis their North European competitors. Moreover, small and medium-sized enterprises, which make up for the largest part of the industrial structure in Southern European countries, often find it difficult, if not impossible, to cope with the costs, particularly in the absence of public subsidies. Finally, the South has no environmental industry, which produces green technology and would thus have financial interests in strengthening environmental standards. Unlike in Northern European countries, EU environmental policies are unlikely to find allies among economic actors in the South.

The rather closed political opportunity structures of the southern member states and their lower level of socio-economic development tend to discourage societal mobilization in favour of environmental protection, and induce forms of mobilization, which are less effective in pulling EU environmental policies down to the domestic level, respectively. A comparative study of an international research team on several Northern and Southern European countries indicates that the level of environmental activism in Southern Europe may not necessarily be lower than in the North (Kousis, della Porta, and Jimenez 2001). But social mobilization takes different forms; it tends to be more localized, community-based, and *ad hoc*. Instead of formal, centralized and professional environmental organizations, we find loosely structured networks of local environmental groups, which fight against environmentally damaging activities within their neighbourhoods. Environmentalists are able to mobilize local publics by appealing to their strong territorial attachment, which is part of the tradition in southern countries (Weale et al. 2000: 244-246; cf. Kousis, della Porta, and Jimenez 2001; Eder and Kousis 2001).

While environmental organizations have been growing significantly, not least thanks to EU funding, their organizational strength in terms of professional staff, donating members, and budget is still weak compared to their northern counterparts. The most resourceful environmental organizations are transnational NGOs, such as Greenpeace, WWF or Friends of the Earth. Local NGOs can generate pressure on public authorities and economic actors. Yet, such pressure resonates less with the logic of the Pull-and-Push Model, since local groups simply lack the necessary resources to systematically mobilize at the domestic and the European level. Lobbying policy makers in the process of incorporating EU policies into national law requires the money, staff, and expertise of professional pressure groups in order to be effective. So do administrative appeals and litigation before national courts against implementation failures. Not only is litigation time-consuming and resource-intensive. Many environmentalists still distrust their courts as being biased towards public administration and economic interests. At the same time, locally operating groups benefit less from the European opportunity structure. The European Commission is mostly interested in problems of legal implementation, where its compliance pressure works best. Local groups, by contrast, are usually involved in issues of practical application, where they often lack the necessary knowledge and expertise to denounce environmental harmful activities as infringements of European Law. But even if domestic mobilization manages to revoke an instance of flawed or failed application, the impact is limited since their 'victory' is unlikely to affect overall administrative practice. Generating political and legal pressure on policy-makers to correctly incorporate EU environmental Directives into national law (transposition), including the enactment of administrative provisions for practical application, is a more efficient pulling-down strategy. If they have the necessary resources, it allows environmental interests to concentrate their efforts on a limited set of actors during a limited time period. Moreover, the Commission is likely to become a powerful ally exerting additional pressure from above. If the combined pressure from below and from above manages to ensure a correct and complete transposition of EU environmental Directives, this also lays the ground for effective application and enforcement.

To sum up, the Pull-and-Push Model does not deny that the South faces significant problems in complying with EU Environmental Law. But instead of attributing these problems to some generic deficiencies of their domestic institutions or civic cultures, it points to the interaction between the dynamics of EU policy-making and the domestic structures of the southern member states as the major cause of their compliance failures (see figure 3.2). The 'leader-laggard' dynamics of EU environmental policy-making result in significantly higher policy misfit, and hence higher compliance costs for the southern laggards than the northern leaders. As environmental late-comers, southern member states are policy-takers rather than policy-makers. They have neither the incentive nor the capacities to push for environmental harmonization at the European level. As industrial and democratic late-comers, the southern countries not only have weaker compliance capacities in terms of the necessary resources to effectively implement and apply environmental regulations. Their closed political opportunity structures and lower level of socio-economic development discourage societal actors from mobilizing against attempts of public and economic actors to avoid the costs of compliance with European policies.

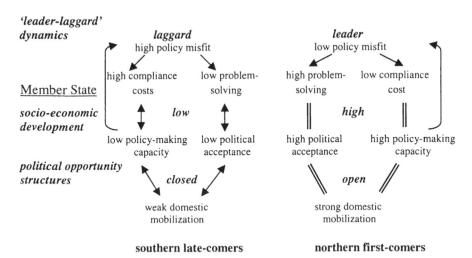

Figure 3.2 The 'Southern Problem' reformulated

The Pull-and-Push Model draws on the existing literature dealing with implementation failure in the South of Europe but manages to overcome two major deficiencies. First, it avoids the reductionism of the 'Mediterranean Syndrome' approach by pointing to explanatory factors that are neither geographically bound nor static.

The capacity to upload domestic policies to the European level, on the one hand, and the openness of political opportunity structures as well as the level of socio-economic development, on the other, not only vary across member states, both in the North and the South of Europe. They may change over time, which means that the Southern Europeans are not doomed to be environmental laggards but are able to catch up with the northern leaders, particularly if EU policy-making becomes more responsive to their environmental problems and economic concerns. Second, the Pull-and-Push Model systematically links policy and institutional factors at the European and the domestic level in explaining different levels of compliance. It thereby allows to account for both cross-country and cross-policy variation. The next chapter presents 12 case studies on the implementation of six EU environmental policies in Spain and Germany, which systematically test the explanatory power of the model.

Chapter Four

Implementing EU Environmental Policies in Germany and Spain

Implementation failure and non-compliance in the EU are not a genuine 'Southern Problem' caused by a disease from which only the Mediterranean member states suffer. While it is true that infringements of European Environmental Law are more prevalent in the South than in the North, they result from a constellation of EU- and domestic factors that render effective implementation and compliance more difficult for southern EU-policy takers than for northern EU-policy makers. In order to state the argument, this chapter compares the implementation of six different EU environmental policies in Germany, one of the environmental leaders, and in Spain, which pertains to the group of environmental laggards. The chapter starts with a short explanation of why the six policies were selected and how effective implementation and compliance are measured. Then, the major features of environmental policy-making in Germany and Spain are briefly described. The main part of the chapter traces the process through which each of the policies is implemented in the two member states. The comparison shows that strong policy misfit gives rise to serious implementation problems for leaders and laggards alike. Furthermore, compliance in both countries significantly improves when domestic actors and the European Commission pressure public authorities and policy-makers from below and from above. The study not only testifies that environmental leaders may run into significant compliance problems, too, but demonstrates that the same factors promote and impair effective implementation and compliance in the member states, irrespective of their geographical location or their political culture.

Selecting the Cases: High Policy Misfit and Differing Degrees of Domestic Mobilization

Policy Misfit as the Necessary Condition for Non-Compliance and Implementation Failure

The comparative study looks at the implementation of the following six EU environmental policies:

- the Drinking Water Directive;
- the Directive on the Combating of Air Pollution of Industrial Plants;

- the Large Combustion Plant Directive;
- the Environmental Impact Assessment Directive;
- the Access to Information Directive;
- the Eco-Audit Regulation.

The first three Directives on Drinking Water and Air Pollution Control are examples of traditional policies entailing a technocratic and interventionist approach of top-down regulation which imposes uniform and detailed standards on the member states. The latter three policies on Environmental Impact Assessment, Access to Environmental Information, and Eco-Audit Management Systems, by contrast, represent the 'New Policy Instruments' (NPI), which emerged with a reorientation in the mode of European governance towards network-style and bottom-up forms of environmental policy-making in the early 1990s (Knill and Lenschow 2000b). NPIs emphasize regulation that is cross-media and procedural, and builds on public participation and voluntary agreements with regulated parties to ensure compliance. Given their 'new' approach, NPIs may cause serious policy misfit even for environmental first-comers, like Germany, whose regulatory structures are firmly rooted in a traditional approach of substantive, top-down regulation. Thus, the European policies on Environmental Impact Assessment, Access to Environmental Information, and Eco-Audit Management Systems are likely to cause significant compliance problems for Spain and Germany alike. In fact, all six policies impose serious adaptational costs for both countries, except for the two Air Pollution Control Directives in the case of Germany.

Differing Degrees of Pull-and-Push

While policy misfit is kept constant for ten of the 12 cases, the level of 'pull and push' varies significantly; Germany and Spain face different degrees of internal and external pressure for adaptation in the implementation process. Given its lower level of socio-economic development and its relatively closed political opportunity structure in environmental policy-making (see below), adaptational pressures should be lower in Spain than in Germany. Indeed, in three of the six Spanish misfit cases – the Drinking Water Directive, the Industrial Plant Directive, and the Eco-Audit Regulation – neither domestic actors nor the Commission have become active, despite serious implementation problems regarding the first two Directives. German domestic actors, by contrast, mobilized in all but one of the four misfit cases (Drinking Water). The same is true for the Commission, which only remained inactive in one case (Eco-Audit), where non-compliance is less an issue given the voluntary nature of the Audit and Management Scheme. For the Environmental Impact Assessment and the Access to Information Directives, however, both Spain and Germany have felt significant pressure from below as well as from above. The remaining cases fall in-between the two opposite poles (no vs. combined pressure). The Large Combustion Plant in Spain and the Eco-Audit Regulation in Germany gave rise to domestic mobilization of societal and economic actors, respectively, while the Commission has remained inactive. For the Drinking Water Directive in Germany, the situation is the other way around – while the Commission has exerted considerable pressure on German authorities, domestic actors have largely refrained from mobilizing. The Air Pollution Control Directives for Germany are left aside here since the two policies have not produced any misfits and hence do not fulfil the necessary condition for implementation failure and non-compliance.

Table 4   Ten cases of policy misfit and differing degrees of pull-and-push

Pressure from Below	*Pressure from Above*	*Commission active 'Push'*	*Commission inactive*
Domestic actors mobilizing 'Pull'		Access to Information Germany and Spain	Large Combustion Plant Spain
		Environmental Impact Assessment Germany and Spain	Eco-Audit Germany
No domestic mobilization		Drinking Water Germany	Drinking Water Spain
			Industrial Plant Spain
			Eco-Audit Spain

If the assumptions of the Pull-and-Push Model hold, we shall expect:

- *full compliance* with the Industrial Plant and the Large Combustion Plant Directives in Germany;
- *sustained non-compliance* with the Drinking Water Directive, the Industrial Plant Directive, and the Eco-Audit Regulation in Spain;
- *emerging compliance* with the Directives on Access to Information and Environmental Impact Assessment in Germany and Spain, the Drinking Water Directive and the Eco-Audit Regulation in Germany, and the Large Combustion Plant Directive in Spain.

Measuring Effective Implementation and Compliance

How to Define Effective Implementation and Compliance?

The literature presents a Babylonian variety of understandings and definitions of compliance and effective implementation. Moreover, the distinction between the two concepts is often not clear. For the sake of this study, compliance is defined as *rule-consistent behaviour of those actors, to whom a rule is formally addressed and whose behaviour is targeted by the rule.*[1] While states are the addressees of European Law, and, hence, are formally responsible for compliance, they are not necessarily the main or exclusive targets. Many policies target the behaviour of non-state actors, too. If states are only the addressees but not the main targets, the ultimate responsibility for compliance stays with private actors. Beyond formal incorporation into national law, the role of public actors is confined in these cases to effectively monitoring and enforcing European policies in order to ensure compliance (Börzel 2002a).

The distinction between addressees and targets of a rule or policy helps to clarify the relationship between *compliance* on one hand, and *implementation*, on the other hand. *Implementation* refers to the putting into practice of policies or rules. Drawing on David Easton's system theory approach, implementation studies often distinguish between three different stages of the implementation process:

- *output*: the legal and administrative measures to put a policy into practice (formal and practical implementation);
- *outcome*: the effect of the policy measures on the behaviour of the target actors;
- *impact*: the effect of the policy on the socio-economic environment (*effectiveness, problem-solving capacity*).

Compliance defined as rule-consistent behaviour of both the addressees and the targets of a rule or policy comprises the output and the outcome dimension. Impact and effectiveness are a separate matter since compliance of rule addressees and rule targets need not lead to changes in the socio-economic environment. This study concentrates on output and only considers outcome as far as the compliant behaviour of state actors is concerned. The omission of the rule-consistent behaviour of non-state actors appears to be justified for three reasons. First, the European legal system is geared towards the member states. They are exclusively responsible for the effective implementation of EU policies at the domestic level (Weiler 1988; Krislov, Ehlermann, and Weiler 1986). Even in cases where they are only the addressees and not the major targets of a policy, the EU holds the member states

---

[1]   Actors targeted by a policy are not identical with actors affected by a policy. Unlike affected actors, a policy directly requires target actors to change their behaviour. For instance, consumers are affected by foodstuff regulations, but it is the food industry, which has to change its production behaviour.

responsible for non-compliance. European infringement proceedings pursue only indirectly non-compliant behaviour of private actors by reprimanding the member states for not effectively monitoring and enforcing European Law. Second, the Pull-and-Push Model follows the logic of European Law and focuses on the non-compliant behaviour of state actors targeted by domestic actors and the Commission. Pressure from below and from above shall 'persuade' policy makers and public authorities to confront the costs of implementing 'misfitting' policies and enforce them on recalcitrant private actors. Third, analyzing the implementation of six different EU policies in two member states requires considerable resources. It would have gone beyond the scope of this research project to also examine the rule-consistent behaviour of private actors in 12 cases, some of which span across a period of more than 20 years.

How to Measure Effective Implementation and Compliance?

If we confine our dependent variable to the rule-consistent behaviour of state actors, a policy is considered as effectively implemented and complied with if:

- the policy is completely and correctly incorporated into national legislation and conflicting national rules are amended or repealed (*formal implementation*);
- the administrative infrastructure and resources are provided to put the objectives of the policy into practice and to monitor the rule-consistent behaviour of the target actors (*practical application*);
- the competent authorities encourage or compel rule-consistent behaviour of the target actors by effective monitoring, positive and negative sanctions, and compulsory corrective measures (*monitoring and enforcement*).

The empirical study in this chapter assesses the implementation of six EU environmental policies in Germany and Spain. The assessment evolves along several steps. First, a careful analysis of the policy content as codified in the Directive or Regulation shall determine the requirements, which the member states have to fulfil. Which is the problem the policy shall address? What are the solutions offered (policy goals)? Which is the dominant problem-solving approach on which the policy is based? Which instruments were chosen to achieve the policy goals? Most importantly, which are the concrete obligations posed to the member states? The analysis will also look at the 'history' of the policy in order to identify patterns of the leader-laggard dynamics in EU policy-making.

In the next step, the fit between the EU policy and already existing policies in the two member states is assessed in order to determine the compliance costs. Have Germany and Spain already policy structures in place, which tackle the issue, addressed by the EU policy? To what extent are domestic problem-solving approaches, policy instruments, and policy standards compatible with those prescribed by the EU policy? Which adaptations are necessary to integrate the EU policy into existing legal and administrative structures? What resources are required to ensure the effective application, monitoring and enforcement of the policy?

The third step analyzes the implementation of the policy in Germany and Spain. How was the policy formally implemented? Which legal and administrative measures have been enacted and modified to integrate the policy into national legislation and to put it into practice? Which actors, public and private, tried to interfere with the formulation and decision-making of the domestic legislation transposing the EU policy into national law? Which resources were allocated to ensure the application, monitoring and enforcement of the policy? What kind of actions do public authorities take to monitor compliance with the policy? How do they respond to instances of non-compliance?

The fourth step assesses the extent to which the two member states have effectively implemented and complied with the EU policy. The legal and administrative output and implementation behaviour of public actors observed under the third step are compared to the requirements of the policy established under the first step. Are the legal and administrative adaptations sufficient to satisfy the requirements of the EU policy? Have human resources, financial means, expertise, and information been made available to ensure the effective application, monitoring, and enforcement of the policy? Have public authorities systematically sanctioned policy violations?

In cases of implementation and compliance failure, the next step seeks to reveal the causes of non-compliance. Have powerful domestic actors blocked the political decisions necessary to formally implement the EU policy? Do public authorities have sufficient resources to practically apply, monitor and enforce the policy? How do public authorities justify the infringements? Do they accept the EU policy as legitimate?

Finally, the pull-and-push mechanisms are examined. Are domestic actors aware of implementation and compliance failures? How do they respond to instances of non-compliance? Do they take political or legal action? Do they seek to mobilize the Commission? Has the Commission introduced infringement proceedings? Do public authorities and policy-makers yield to pressure from below and from above? Does compliance improve over time? Are there factors other than external and internal pressure for adaptation, which help to reduce instances of non-compliance? Have European institutions provided resources to help a member state coping with the implementation costs? How important are intergovernmental and transnational networks in which public and private actors exchange information and experience about the implementation of EU policies?

Assessing whether a policy is effectively implemented and complied with is not an easy task. A judgement is never objective since it depends on the researcher's interpretation and evaluation of the data collected. Therefore, the study seeks to back its judgements by as many sources as possible, both critical and sympathetic to the addressees and targets of the EU policies. Next to secondary literature and legal and policy documents, the empirical analysis heavily draws on interviews with legal and policy experts, environmental groups, European and domestic policy-makers, implementers (competent authorities), and some of the major targets of the policies under consideration (e.g. companies).[2]

Environmental Policy-Making in Germany and Spain

Germany as an Environmental Policy-Maker and Pace-Setter

Policy principles and policy style State regulation to protect humans and their environment against hazardous industrial activities has a long tradition in Germany (Pehle and Jansen 1998: 84-86; Wey 1982). Already in the 19th century, Germany had enacted several pieces of environmental legislation, such as the Prussian Industrial Statute of 1845. But it was not until 1969 that environmental policy developed into a national policy area based on a comprehensive concept of environmental protection (Hartkopf and Bohne 1983). The new centre-left coalition government under Chancellor Willy Brandt, which came into power in 1969, gave high priority to environmental protection and nature conservation and enacted various environmental programmes, a series of new environmental laws[3] and a number of institutional changes such the creation of the Council of Environmental Advisers (*Rat der Sachverständigen für Umweltfragen/SRU)*[4] in 1971 and the Federal Environmental Office (*Umweltbundesamt/UBA)*[5] in 1974 (cf. Wey 1982; Müller 1986). Initially not having been an environmental forerunner, Germany

---

2   For a complete list of the more than 100 interviews conducted see Börzel 1999.

3   New federal legislation included the Air Traffic Noise Act (1971), the Leaded Petrol Act (1972), the Waste Disposal Act (1972), the DDT Act (1972), the Federal Air Pollution Control Act (1974) and the Environmental Statistics Act (1974) as well as a number of regulations, decrees and administrative directives, such as the Administrative Directive on Clean Air (*TA Luft*) of 1974. After a period of stagnation from about 1974 till 1978, mainly due to the oil crisis, further regulations were passed such as the Nature Conservation Act (1976), the Waste Water Charges Act (1976), The Amendment of the Water Resources Management Law (1976), the Chemical Act (1980), the Ordinance on Large Combustion Plants (1983), the Waste Avoidance and Management Act (1980), cf. Pehle 1997.

4   The Council of Environmental Advisers is made up by independent social and natural scientists. Its main purpose is to advise the government on environmental matters.

5   The Federal Environmental Office was modeled after the US Environmental Protection Agency. Unlike its American counterpart, however, it has less autonomy and no regulatory functions. Its main task is to provide planning, documentation and information as well as technical, scientific and administrative support to the Federal Ministry of Environment. The *UBA* takes responsibility for researching standards and drafting potential regulations in a way that is legally and constitutionally sound. It also coordinates the implementation of environmental laws as well as the collection and aggregation of data at the *Länder* level (Weale et al. 2000: 207-208).

became a leader in environmental policy, making and shaping environmental policies both at the domestic and at the international level (Weidner 1995).

German environmental legislation is a historically grown collection of medium-specific, often highly detailed regulations at the federal and regional (*Länder*) level. It constitutes one of the most complex and densely regulated legal systems of environmental protection in the world (Jänicke and Weidner 1997: 138). There are more than 800 environmental laws, some 3000 environmental ordinances and around 4800 administrative directives. A cross-sectoral integration of the various medium-specific regulations is still wanting (SRU 1994: 177). Since 1986, repeated attempts have been made to reduce the complexity and increase the consistency of environmental legislation. Several draft environmental codes have been published, the first in 1973, and the most recent in 1997 (Bundesministerium für Umwelt 1997), but none has succeeded. While an amendment of the German Constitution in 1993 made environmental protection a state duty, Germany still lacks an Environmental Code. The large majority of environmental policies follow a command-and-control approach, which is based on the traditional police law with its emphasis on danger avoidance (*Gefahrenabwehr*). It corresponds to a statist and legalist tradition, according to which the state hierarchically intervenes in society on the basis of the law and which seeks to remove the area of administrative decision-making from the partiality of political interests (Dyson 1980: 9). German administration is to apply legal standards in the context of formalized procedures (going by the books). This rather rigid approach is difficult to combine with more flexible instruments, which draw on the setting of quality standards, self-regulation, and public participation rather than on stringent and detailed prescriptions and prohibitions (*Gebote und Verbote*).[6]

While Germany still lacks a general framework for environmental legislation, there is a clear statement of three major policy principles, in which German environmental protection is firmly anchored: precaution (*Vorsorgeprinzip*), the polluter-pays (*Verursacherprinzip*) and cooperation (*Kooperationsprinzip*).

The precautionary principle stresses the advantages of prevention over remedial action in environmental policy. It is based on the minimization of environmental hazards and the sustainable and careful use of resources. This even applies in the absence of sufficient evidence to justify the claim that a particular type of activity is harmful to the environment. As such the precautionary principle is directed against end-of-the-pipe solutions calling for policies which fight the source of environmentally harmful effects. The strict application of the best clean-up technology available irrespective of economic costs (*Stand der Technik*) as well as the emphasis on emission control over an ambient quality approach are major elements of the precautionary principle.

---

6   Yet, the capacity of German authorities to implement rigid legal standards strongly depends on patterns of informal and cooperative bargaining between public authorities and regulated parties, to which the literature has referred to as 'cooperative administration' (*kooperative Verwaltung*, Benz 1984) or 'informal rule of law' (*informeller Rechtsstaat*, Bohne 1981).

The polluter-pays principle is directed against the externalization of environmental costs. The costs of preventing, eliminating and correcting environmental harmful effects are to be borne by those who are responsible for the effects in the first place. To internalize environmental costs, German environmental policy heavily relies on traditional instruments of command-and-control, such as prescriptions and prohibitions (Cremer and Fishan 1998). Economic instruments inspired by the polluter-pays-principle (taxes, charges, tradable permits) play only a minor role in environmental protection. They are mainly used in the areas of waste water (effluent charges), petrol, and waste management. But the efficiency and effectiveness of these instruments is rather low (SRU 1994).

The cooperation principle, finally, refers to the joint responsibility of state and society for environmental protection. It aims at the participation of all relevant actors in the environmental policy-making process. Environmental policy-making in Germany is indeed characterized by an intensive formal and informal cooperation between public and private actors. A network of advisory bodies, expert groups, and consultative committees as well as informal relationships ensure the participation of industry and experts in both policy formulation and policy implementation (Aguilar Fernández 1997).

Policy-making competencies Initially, the Basic Law had not assigned the Federal State (*Bund*) much responsibility in the area of environmental policy leaving it in the responsibility of the *Länder*. In the late 1960s, however, the new centre-left coalition government of Willy Brandt pushed for comprehensive federal competencies it deemed necessary for realizing its ambitious environmental agenda. In 1972, a constitutional amendment granted the *Bund* concurrent legislative power for statutory regulations on waste management, air pollution control, noise abatement, protection from radiation, and on criminal law relating to environmental protection matters (Art. 74 GG, Nr. 24). Since then, the *Länder* have only been able to legislate on these issues unless and until the *Bund* takes on responsibility. In the areas of water resources, waste management, regional planning, nature conservation, and landscape preservation, the *Länder* were only willing to give the *Bund* competencies for framework legislation (Art. 75 GG), which would leave it to them to provide detailed and specific legislation (cf. Müller 1986).

In spite of the subsequent centralization of environmental competencies, the *Länder* have substantially shaped environmental policies in Germany. Their participation in making federal laws through the *Bundesrat*, the Second Chamber of the federal legislature, grants the *Länder* significant influence on the formulation of federal regulations. The *Bund* may have the power to regulate all important fields of the environment. It usually transposes EU environmental policies, too. But the federal government needs the consent of the *Bundesrat* in most of the cases. Moreover, it depends on the *Länder* for the implementation of its environmental laws given the limited size of the federal administration. This functional distribution of labour between the central and the regional level of government have fostered comprehensive patterns of joint decision-making and interlocking politics (*Politik-*

verflechtung), which are characteristic for German cooperative federalism in general.

The Standing Conference of Environmental Ministers (*Umweltministerkonferenz/UMK*) and the environmental committee of the *Bundesrat* are the two major axes of vertical and horizontal intergovernmental coordination in environmental policy-making. The *UMK* was established in 1973 in order to provide a closer cooperation between *Bund* and *Länder* in environmental policy-making. It convenes at least twice per year and deals with questions of the *Bund* and *Länder* legislation, the enactment of ordinances and administrative directives and problems with the implementation of environmental legislation. The core of the system of joint decision-making in environmental policy form the various *Länder* and *Bund-Länder* committees and working groups, which exist in the different environmental sectors. Even in those areas, in which the *Länder* may still legislate themselves, they seek to agree on common environmental standards rather than pursue unilateral initiatives. The establishment of homogeneous (federal) policies conforms to a consensus-oriented political culture, which reflects a strong preference for uniform and equal living conditions to be enjoyed by all Germans irrespective in which part of the country they live (Böckenförde 1980). The formal and informal institutions of joint decision-making and policy coordination have largely prevented any substantial regulatory competition among the *Länder* and help to ensure a uniform implementation of EU and national environmental law (cf. Börzel 2002b: chapter 4). But they also provide the *Länder* with a powerful veto position in federal environmental policy-making, which is often blamed for Germany's failures to effectively implement European Law. The *Länder* have indeed repeatedly blocked and delayed the implementation of costly EU policies. At the same, they are only one, albeit powerful veto point in the implementation process. The corporatist patterns of interest intermediation grant interest groups, such as business and farmers, with similar opportunities to impair the effective implementation of 'misfitting' EU policies (see below).

According to the constitutional right of self-administration (Art. 28 GG), local authorities also fulfil important environmental tasks either as their own competencies or because they were delegated to them by the *Bund* and the *Länder*. The municipalities execute about 80 per cent of the federal and regional legislation (cf. Müller-Brandeck-Bocquet 1996: 171-180; Weale et al. 2000: 202-209). Unlike the *Länder*, the municipalities have no formal input on federal environmental policy making. Their central association can only lobby federal and regional decision-makers in order to prevent policies, which they perceive as detrimental to their interests (such as the Ordinance on Packaging Waste). Important areas of environmental protection assigned to the local authorities are cleansing and waste sewage, waste recycling, cleaning up contaminated sites, noise abatement, or nature and landscape conservation. The intermediate level between *Länder* and municipalities (*Bezirke, Landkreise*), finally, has an important role in licensing and supervising environmentally harmful activities.

Next to the Federal Ministry of Environment, Nature Conservation and Nuclear Safety (*BMU*), which replaced the Federal Ministry of the Interior as the competent ministry in 1986, there are various other federal ministries with specific responsibilities in environmental policy-making and other environment-related areas, including the Ministry of Economics, the Ministry of Transport, the Ministry of Agriculture, the Ministry of Building and Planning, and the Ministry of Education, Science, Research and Technology. Given these powerful rivals, the Ministry of Environment has a rather weak position in setting the agenda and formulating environmental policies. This is particularly true at the European level, where other ministries often seek to dominate the formulation and representation of the German bargaining position on environmental issues (Pehle 1997: 198-199). Nevertheless, the *BMU* has been able to provide for a relatively high concentration of environment functions, which has helped to counteract problems of horizontal fragmentation (Weale et al. 1996).[7]

At the regional level, the situation is similar. Bavaria was the first to establish a Ministry of Regional Development and Environmental Affairs in 1970. By now, all the 16 *Länder* have a ministry mainly or exclusively responsible for environmental matters. The *Länder* Ministries of Environment largely follow the functional task structure of the *BMU* (cf. Smollich 1992).

Patterns of interest intermediation   There is a strong corporatist tradition of institutionalized cooperation between government and industry, regulator and operator (Héritier, Knill, and Mingers 1996: 59). Public authorities also maintain informal relations with regulated parties in order to ensure the effective application of rigid legal standards. Such patterns of 'cooperative administration' (Benz 1984; Voigt 1995) grant economic interests privileged access to public policy-making. Likewise, the powerful German Farmers' Union (*Bauernverband*) has exercised significant influence on environmental policy-making. Public authorities also rely on the technical and scientific expertise of private actors in the formulation and implementation of environmental policies. For instance, technical standards often draw on regulations developed by non-governmental organizations including the German Standards Institute (*Deutsches Institut für Normung/DIN*) or the Association of German Engineers (*Verein Deutscher Ingenieure*).

Compared to industry, farmers and scientific experts, environmental interests have still limited access to and influence on the policy process (Hey and Brändle 1992: 77-78). Citizen participation in environmental matters has a long tradition in German administrative practice. It has evolved, in particular, in the context of local land use and construction planning and has been further codified in planning permit procedures (*Planfeststellungsverfahren*) and administrative licensing procedures. Yet, in most cases, citizen participation concerns preventive measures only and is often reduced to ex-post information on plans and decisions already negotiated and agreed between investors and the authorizing administration. Thus, participation tends to 'degenerate' to a confrontational exchange of views instead of encouraging communication and cooperation.

---

[7]   For a more skeptical view see Müller 1994.

Organized societies for nature conservation go back to the 19th century while new types of environmental organizations emerged in the 1970s. Parallel to the anti-nuclear movement, citizens' initiatives (*Bürgerbewegungen*) formed as a result of tensions between citizens and planning authorities at the local level. They organized themselves at the federal level in the Federation of Citizens' Groups for Environmental Protection (*Bundesverband Bürgerinitiativen Umweltschutz/BBU*). The groups gained political importance through the mobilization of candidates in elections and played a major role for the breakthrough of the Greens into the party system in the 1980s, as a result of which established parties became more responsive to environmental issues (Pehle and Jansen 1998: 87-92). Alongside the growing number of citizens' initiatives, there are a large variety of local, regional, and national environmental organizations. They have around four million members (about 5 per cent of the German population, cf. Weale et al. 2000: 258).

Despite their organizational strength and the support they can draw from the high level of environmental awareness among the German public, representatives of environmental interests often remain outsiders to the policy process. For them, the public administration is still a rather closed system, which imposes environmental policy in a top-down manner. Policy-makers and administrators in turn tend to look on environmental organizations as opponents rather than partners. The ministerial bureaucracy considers environmentalists to be 'incompetent, unable to reach compromises, and too 'radical' in the sense that they support unrealistic and non-realistic demands' Pehle 1997: 177). The ongoing professionalization of the environmental groups has led to certain improvements in the relationship with both public authorities and industry (Jänicke and Weidner 1997). The latter have come to accept the expertise and legitimacy, which environmental groups may offer when embracing a more lobbyist approach in their political interaction (Pehle 1997: 178). Accessibility for environmental groups has also improved since the Green Party entered a coalition with the social democrats to form the federal government in 1998. But while environmental groups have increased their public support both in terms of members and contributions and managed to broaden their access to the policy process, the environmental lobby remains relatively weak. The financial base of most environmental organizations is insufficient to ensure systematic input to policy design and monitoring. Voluntary contributions can be obtained more easily for spectacular actions and projects with high visibility. Moreover, environmental organizations have no standing in law suits (*Verbandsklagerecht*) as representatives of the common interest.[8] Another reason for the limited capacity of environmental interests to influence policy formulation and implementation is that the German environmental movement is highly fragmented

---

[8]   The Federal Nature Conservation Act establishes participatory rights for recognized environmental organizations, but does not foresee legal standing. Most *Länder* provide opportunities for NGO to seek legal action. However, some of the larger *Länder*, including Baden-Württemberg, Bavaria and North-Rhine-Westphalia, have no such provisions. Here, NGOs can only work through affected citizens who are willing to bring their case before the courts.

and different groups often act as rivals rather than allies in the policy process. Not only do they compete for resources (members, funding). Nature conservation groups, such as the German Nature Conservation Association (*Deutscher Naturschutzbund*), often pursue different interests from ecological organizations, such as Greenpeace, like in case of wind power plants, which the latter favour as alternative energy sources, while the former denounce them as a disturbance to the landscape.

All in all, environmental policy-making in Germany can be characterized by a legalistic and interventionist but consensual policy style. It entails a precautionary problem-solving approach favouring prevention over reduction of environmental pollution and considering uniform emission limit values (as opposed to quality standards) and the best available technology, regardless of its economic costs, as the most appropriate means of environmental protection.

Preferences and strategies in EU policy-making Germany is a country, which favours stringent environmental regulations. At the same time, its economy is highly export oriented. Thus, Germany has a strong incentive to harmonize its policies at the European level in order to reduce institutional adaptation costs and avoid competitive disadvantages for its industry, respectively.

In the 1980s, Germany was quite successful as a maker and shaper of European environmental policy exporting its stringent substantive regulations to other member states via European harmonization, especially in the field of water and air pollution control. Germany has both the policies and the resources to act as an active pace-setter of European environmental policies. In the late 1980s, however, German enthusiasm for functioning as 'the motor behind EU environmental policy' (Liefferink and Andersen 1998a: 71) started to cool off. On the one hand, unification imposed a considerable economic burden on Germany. Sluggish economic growth and high unemployment led to a certain shift in German priorities and have dampened public and government ambitions for increased environmental protection (Pehle 1997). Germans are less concerned about environmental protection than they were in the early 1990s, and their willingness to pay for environmental protection measures has decreased substantially (OECD 2001: 152). On the other hand, the 'new politics of pollution' emphasized the need to integrate environmental measures with other sectors of public policy (Weale 1992). The integrated approach has posed significant difficulties for Germany in incorporating European environmental policies into its highly sectorized regulatory structure. Moreover, European procedural measures, which seek to enhance public participation and encourage voluntary self-regulation by economic actors, are at odds with the traditional regulatory approach in Germany where the state takes an authoritative role in setting environmental standards and defines how they are to be implemented. As a result of the paradigm shift in European environmental policy-making (cf. Heinelt et al. 2001), Germany has found itself increasingly in the role of a taker rather than a maker of European policies, including Environmental Impact Assessment and the Integrated Pollution Prevention and Control Directive. Since these policy innovations have incurred considerable adaptational costs at the domestic level, Germany has repeatedly defected from the leader group, which also shows in its implementation record.

> Germany has been left behind, so to say; it is trying to prevent, or at least delay, the new regulatory philosophy from being put into concrete terms by the Council of Ministers and, thus, it implements the decisions taken in Brussels into national law only haltingly, insufficiently, and frequently with delay (Pehle 1997: 201-202).

The distinction, which German bureaucrats proudly made in the 1980s between *"die Germanen"* ... who play it straight on detailed implementation, and *"die Romanen"* ... whose bureaucracies often frustrate the goals of European policy at the stage of detailed implementation' (Bulmer and Paterson 1987: 183) does not hold anymore.

Spain as an Environmental Policy Taker and Foot-Dragger

Policy principles and policy style Environmental policy has not been a priority on the political agenda of Spanish policy-makers. The few regulations predating EC accession had been mostly concerned with the protection of human health from harmful industrial activities at the local level (Martin Mateo 1977; Font and Morata 1998: 211-212). For instance, the national regulation 'Unhealthy, Harmful and Dangerous Activities' of 30 November 1961,[9] prescribes a licensing procedure for the authorization of classified activities to be done by the municipalities. Likewise, the Air Pollution Act of 1972,[10] which ranks among the first modern pieces of environmental legislation in Spain, establishes a basic framework for air pollution control (see below). With the transition to democracy in 1975, the Spanish administration became more active in environmental policy-making, especially in the sector of water pollution and hazardous waste. The Water Act (*Ley de Aguas*) of 1986 modernized the Spanish water legislation dating back to 1879 and adapted it to European standards. But in general, policy-makers were more concerned with the consolidation of democratic institutions and the economic development of the country than with building-up a regulatory structure for fighting environmental pollution. Belonging to the periphery of Europe, economic growth and employment have been the overall political priority. When Spain entered the European Community in 1986, its environmental policy was not very well advanced (cf. Aguilar Fernández 1997: chapter 3).

It is difficult to identify particular guiding principles that have motivated Spanish environmental policy-making. Environmental policies are not anchored in a specific problem-solving approach to environmental protection. Rather, environ-

---

[9] *Reglamento de Actividades Molestas, Insalubres y Peligrosos* (RAMINP), *Decreto* 2414/1961; BOE N° 292, 7.12.1961.

[10] *Ley 38/72, de 22 de deciembre de 1972, de Protección del Ambiente Atmosférico*, BOE N° 309, 16.12.1972.

mental policies have evolved through laws and regulations addressing specific and urgent problems. They primarily rely on instruments of command-and-control regulation. More market-oriented or communicative instruments are hardly employed. Substantial criteria, such as emissions or quality standards, are legally binding and usually do not leave much flexibility in implementation. State action on the environment as in general is based on 'government by decree'. Policies tend to be made by *real decretos leyes*[11] or *real decretos legislativos*[12] rather than by *leyes ordinarias* (ordinary laws to be decided in Parliament). The formalistic and interventionist approach very much reflects a state tradition, which has historically been heavily regulating and interfering in most sectors of social life, and which has been shaped by the particular evolution of capitalism in Spain with its emphasis on state protectionism and almost 40 years of authoritarian rule under Franco (Aguilar Fernandez 1992: 149-152).

Policy-making competencies Like in Germany, the central state shares the responsibility for environmental policy-making with its regions, the Autonomous Communities. The central state is entitled to establish general plans and to set basic legislation (*legislación básica*) in order to provide a lowest common denominator of environmental standards throughout the country. European Environmental Laws are therefore transposed at the national level, whereas practical application and enforcement lie within the responsibility of the Autonomous Communities and the municipalities. The Autonomous Communities are not only in charge of executing basic legislation. They may also enact additional norms, which complete or reinforce the regulations of the basic legislation, for instance by setting stricter standards (Art. 149.23; 148.9 CE).[13] The Autonomous Communities have not equally invoked their competencies in environmental policy. Few have enacted additional regulations that would go beyond central state legislation. Most of the regions simply apply national laws. Only Catalonia has systematically tried to develop a proper body of environmental legislation by enacting its own sectoral laws, such as

---

[11] In cases of urgency and extreme need, the government has the power to introduce rules which have the force of a law (Art. 86 CE). Such *decretos-leyes* may not affect the organization of the basic institutions of the state, civil rights and liberties or the organization and powers of the Autonomous Communities, and have to be submitted to the parliament for approval and ratification within a period of thirty days.

[12] Parliament can decide that some legislation would be better drafted by the government and delegates its legislative power to the government accordingly. This delegation has to be allowed by a formal law.

[13] Before 1992, the ten Autonomous Communities, which had to take the slow track to political autonomy, had strictly speaking lacked the competence to legally develop environmental basic legislation only being allowed to execute it. But as a result of the jurisdiction of the Constitutional Court, which followed a uniform interpretation of Art. 149.1.23, all Autonomous Communities are able to legally develop central state basic legislation (*STC 64/1982, de 4 de noviembre de 1982* and *170/1989, de 19 de octubre de 1989*).

waste, water, air pollution or nature conservation.[14] In contrast to Germany, the Autonomous Communities do not have a formal veto in central state decision-making. Nor does the central state depend on them for the implementation of its legislation since the Spanish government retains its own field services at the regional level to carry out policies of cross-regional or national interest. The presence of the central state within the Autonomous Communities has given rise to repeated tensions between the two levels of government. While the regions have little say in environmental policy-making at the national level, they have a crucial role in the implementation of (EU) policies. But policy coordination between the national and regional level of government used to be weak to non-existing. The Sectoral Conference for the Environment (*Conferencia Sectorial de Medio Ambiente*), which the central state had established in 1988 to facilitate the implementation of its environmental policies in the regions, only started working in 1994 when European issues systematically entered the agenda (Börzel 2002b: 192-208).

Vertical fragmentation in environmental policy-making has been further enhanced by the considerable competencies of the municipalities. The *Ley Reguladora de las Bases del Régimen Local* (basic law governing local administration) grants the municipalities significant environmental responsibilities in the area of waste collection, waste water treatment, provision of drinking water, and the monitoring of air and water quality standards.[15] The provinces, which build an intermediate level between the regions and the municipalities, take on environmental services for the smaller local communities. But even larger municipalities often lack sufficient technical expertise and financial resources to properly fulfil their environmental responsibilities. Some have tried to overcome the problem by pooling their resources in inter-municipal bodies. This has further increased the complex overlap of environmental competencies between four levels of government, each of which jealously guards its own jurisdiction. The conflict laden triangular relationship between the national, regional and local level of government does not allow it to systematically address this problem but rather reinforces it.[16]

---

[14] Other Autonomous Communities, like Galicia, Andalusia, the Basque Country, the Canary Islands and Madrid, have occasionally followed the example of Catalonia in enacting their own regional laws which do not merely operationalize central state legislation.

[15] All municipalities must organize waste collection, street cleaning, water supply and sewerage. Those with fewer than 5000 inhabitants are encouraged to undertake these tasks jointly (*mancomunidades, entidades metropolitaneas*). Municipalities with more than 5000 inhabitants must manage their own solid waste; those with more than 50,000 inhabitants must have a fully-fledged environmental protection service. The municipalities also issue permits for buildings and certain industrial installations, but are subordinate to the regional governments regarding land use planning and permits (cf. Choy i Tarrés 1992; Perdigó Solà 1996; Font and Morata 1998: 220-221).

[16] The municipalities can ask the Autonomous Communities to suspend their obligation to provide certain services if they lack the capacity to do so (Art. 26.2. *Ley 7/1985, de 2 de abril, Reguladora de las Bases del Régimen Local*). Waste water treatment, an exclusive

Vertical fragmentation in environmental policy-making has been complemented by 'intersectorial discoordination' (Aguilar Fernandez 1992: 50) of environmental responsibilities at the central state level, which are widely dispersed
among different ministries. Horizontal coordination, however, has been improved
by the subsequent up-grading of the administrative unit responsible for the
environment, culminating in the creation of the *Ministerio de Medio Ambiente*
(*MIMA*) in May 1996. The new ministry unites environmental functions previously
allocated among eight ministries and over 20 Directorates General.[17] Its position,
however, is still weak since other ministries have been resistant to give up their
former competencies. In the Autonomous Communities, environmental policy-
making is equally fragmented due to the dispersion of environmental responsibilities and the weak position of environmental authorities, which lack co-ordination
capacity. All regions installed proper environmental authorities, of which only
some enjoy the status of an independent ministry (*Departamento, Consejería*). The
majority of the regions combined environmental responsibilities with other
functions such as economy, culture, agriculture, regional planning or health.

Patterns of interest intermediation   Authoritarian state interventionism kept civil
society weak. The degree of societal self-organization used to be low, and policies

---

competence of the municipalities, is a typical example for the *de facto* substitution of local responsibilities by the regional and central state administration. The municipalities,
however, started to oppose the increasing centralization of their competencies at the regional level. They have called for a *Pacto Local* to strengthen their ability of exercizing
their competencies (basically by providing them with more financial resources) rather
than having them taken away by the Autonomous Communities (cf. Poveda Gomez
1997). The Autonomous Communities are in general reluctant to strengthen the local
level because they fear to lose competencies.

17   Before May 1996, two ministries were primarily responsible for Spain's environmental
policies: the Ministry of Public Works, Transport and Environment and the Ministry of
Agriculture, Fisheries and Food. The Ministry of Public Works, Transport and Environment included a State Secretariat of Environment and Housing, which was responsible for co-coordinating national environmental policies and legislation concerning water, air and waste. The Ministry of Agriculture, Fisheries and Food was in charge for
legislative matters concerning nature conservation and forestry. It had a Directorate
General on Nature Conservation, which was also responsible for the management of national parks and the national register of sites of ecological importance. Other ministries
with important environmental responsibilities are: the Ministry of Industry and Energy,
which is responsible for policies concerning the environmental impact of industry and
for industrial innovation and technology, and is in charge of formulating Spain's energy
policy; the Ministry of Health and Consumption, which is responsible for health monitoring; and the Ministry of the Interior, which is in charge of enforcing laws on water,
nature protection and the prevention of major industrial accidents. To ensure coordination among ministries, interministerial commissions are established for specific policy
issues such as climate change or international environmental cooperation (cf. Ortega Al-
varez1991; Font and Morata 1998: 215-217).

have been made without any substantive participation of societal actors. This remains true also for the time after Spain's transition to democracy. Institutional channels for consultations with the different stakeholders have been largely absent. Economic interests have increasingly enjoyed some informal but discontinuous relations with the public administration, particularly at the implementation stage. Environmental authorities also regularly consult specialist agencies and policy experts. Societal actors, by contrast, are still largely excluded from the policy-making process (cf. Aguilar Fernández 1997). The environmentalist movement only started to form in 1973 mobilizing against nuclear power station projects. While movement organizations were largely characterized by local activism and ideological diversity, some environmental groups became more stable and less radical. They started to finance themselves through membership contributions and public grants. Some of them organized themselves in a national federation, the *Coordinadora de Organizaciones de Defensa Ambiental* (CODA). Together with the Spanish sections of transnational environmental organizations, such as Greenpeace, WWF, or Friends of the Earth, they have gained increasing voice in the political process. In 1994, the Spanish government established the Advisory Council for the Environment (*Consejo Asesor de Medio Ambiente/CAMA*), which for the first time provides some institutionalized participation of societal interests in central state environmental policy-making. *CAMA* represents a variety of interest groups, including business, trade unions, consumers, agriculture, hunters, scientists, and environmentalists. It has the right to make recommendations on policy proposals put forward by the central government and to bring up own policy initiatives. Yet, economic and societal representatives alike agree that *CAMA* has little influence on environmental policy-making.[18] In 1996-97, two environmental organizations, Greenpeace and AEDENAT, walked out of *CAMA*, protesting against its 'window dressing role' as mere symbolic politics to strengthen the public legitimacy of the Minister of Environment. In the absence of a viable Green Party, environmental interests are not effectively represented in the legislative process either. With the assistance of the German Greens, several Green parties emerged in the early 1980s. But they have faced great difficulties in presenting themselves as a single Green list in national and regional elections. Many local and regional groups have resisted attempts of centralization within the Green movement. Electoral support for environmental concerns is still low. While the majority of the Spanish public deems environmental protection necessary, Spaniards are reluctant to support a party consequently prioritizing environmental issues. Consumerist values still prevail. The situation in the Autonomous Communities is not much different. Policy-formulation and decision-making tend to be as closed at the regional level as at the national level. Economic interests maintain contacts with the regional administration on an informal and personal level. While scientific experts (specialist agencies, consul-

---

[18]   Interviews with environmentalists of Greenpeace, AEDENAT, ADENA-WWF, and CODA, and representatives of the Spanish Business Confederation (CEOE), Madrid, 03/97.

tancies) increasingly form part of these closed policy communities, environmentalists have largely remained outsiders. The process becomes more open at the implementation stage. Regional authorities hope to increase the effectiveness of environmental protection measures by cooperating with economic and environmental actors. Some Autonomous Communities established advisory boards and management councils, which include representatives of business associations, trade unions, farmers, and environmental groups. Moreover, environmental authorities have started to negotiate environmental agreements with industry, particularly in heavily polluted areas (Aguilar Fernández 1997: 203-212). The public funding of a wide range of educational and practical activities of environmental groups has helped to establish closer contact between public authorities and environmentalists (Jimenez 1997). But while the institutionalization of environmental policy at the national and the regional level has broadened the political opportunities for environmental demands and served to some extent as a catalyst for political action, access to the policy process is still limited. Legal action of environmentalists is discouraged by high economic costs on the one hand, and the lack of sensibility and technical training of public prosecutors and judges, on the other hand. The judicial system has served to repress rather than encourage environmental activism by charging environmental activists for committing offences during demonstrations, boycotts and other political actions (Jimenez 1997). The new Criminal Code, which was enacted in 1996, significantly expands the sanctions for environmental offences. It is complemented by a revision of administrative law imposing a system of compulsory fines and remedies against infringements of administrative ordinances (Weale et al. 2000: 312-313). But it is too early to say whether the new legislation will empower societal actors in their fight against environmental pollution.

All in all, environmental policy-making in Spain has been characterized by a legalistic, interventionist, and reactive policy style (Font 1996). It tends to evolve through laws and regulations addressing specific and urgent environmental problems and heavily relying on regulatory, command and control instruments. Unlike in Germany, environmental policies are not based on some clear principles derived from a certain problem-solving approach to environmental protection.

Preferences and strategies in EU policy-making When Spain joined the EC in 1986, it committed itself to the adoption of the whole *acquis communautaire* including the complete body of European environmental legislation enacted before 1986. Unlike Greece and Portugal, Spain did not ask for any transitional period or special conditions of application as environmental protection was not considered a key issue in the negotiations. The process of incorporating the whole range of European environmental policies in Spanish legislation took more than two years and caused a severe policy overload for the Spanish administration. In contrast to Germany, the formal implementation of EU environmental law has not so much been a problem of harmonizing a dense and complex set of existing domestic regulations with European requirements but to bring existing regulations up to European standards. Moreover, often completely new regulations had to be enacted which required the building-up of additional administrative capacity. Not only were new administrative units, procedures and technology to be established for the practical application and enforcement of regulations. Administrators often lacked expertise, that is, the technical and scientific knowledge to apply, and assess compliance with environmental regulations. In 1992, Spain faced an 'environmental deficit' of EUR 30.1 million, which would have been necessary to effectively implement EU environmental legislation (*Medio Ambiente en España* 1992). The Spanish government subsequently increased its public spending on environmental policy, from 0.6 per cent of the GDP in 1989 to more than one per cent of the GDP in 1997 (*Información de Medio Ambiente*, N° 48, February 1997). This is still lower than in Northern European countries. But the per capita income in Spain is only 75 per cent of the EU average. Moreover, the Spanish unemployment rate doubles the average unemployment rate in the EU and is the highest among the OECD countries (OECD 1997).

Spain's role as a taker rather than a maker of environmental policies also manifests itself at the European level where the Spanish government pursues a reactive and often defensive strategy in the negotiations. Spain lacks both the resources and the policies to act as a 'pace-setter' in the EU policy-making process. Spain has been complaining about European environmental legislation, which it perceives as oriented towards the ecological problems and economic interest of the northern industrialized countries. But instead of pushing for European policies that are more responsive to environmental problems of the South, such as the struggle against desertification and deforestation, Spain has most of the time played a rather passive role in the policy formulation and decision-making process. Rather than trying to influence the content of a policy itself, the Spanish government has acted as a foot-dragger asking for temporary derogations, exemptions, and side-payments as to facilitate adaptation at the domestic level. Spain's approval of the Political Union in the Maastricht Treaty was literally bought by the establishment of the Cohesion Fund of which Spain received about EUR 7.9 billion during the period of 1993-99 (Font and Morata 1998: 222; cf. Aguilar Fernandez 1992: 329-334). Another example is the five year exemption from emission reductions, which the Spanish government had negotiated for the implementation of the Large Combustion Plant Directive, saved Spain some EUR 4.1 billion (Aguilar Fernández 1997: 108).

Germany and Spain substantially differ in their economic, social and political structures. As an industrial and environmental late-comer, Spain has developed very distinct patterns of environmental policy-making from Germany, which has been one of the leading countries in terms of both economic development and environmental protection. While environmental groups have less access to the German policy process than business and farmers organizations, the political opportunity structure is far more closed for Spanish environmentalists, whose organizational strength is also considerably weaker than their German counterparts. These differences give rise to varying levels of internal and external pressures for adaptation in the two countries. The 12 case studies of the following section explore how the differing degrees of 'pull-and-push' have shaped the implementation of EU environmental policies and resulted in different levels of compliance.

Drinking Water: Spanish Fit and German Misfit?

The Policy and its Development

The Directive on the Quality of Water on Human Consumption (Drinking Water Directive),[19] adopted by the Council on 15 July 1980, is one out of three European Directives regulating water for human consumption.[20] The Drinking Water Directive covers all water attended for direct human consumption (drinking water) and for food production. It has the dual purpose of promoting the free circulation of goods in the European Union and of protecting human health and the environment. The Directive provides protection of groundwater from pollution caused by dangerous substances. It entails an interventionist problem-solving approach imposing 64 legally binding standards to reach a certain level of drinking water quality. The member states must fix values for each of the parameters indicated in Annex I of the Directive, within the scope prescribed by the guiding levels and mandatory values for maximum admissible concentration and minimum required concentration specified.[21] This legalistic approach does not leave much flexibility for the national administrations in implementing the Directive. Member states, however, can derogate from the standards set in Annex I in a number of circumstances, for instance, following emergencies, stemming from the nature and structure of the ground in the area in which the supply in question emanates, or arising from exceptional meteorological conditions.

While it has been argued that the Directive introduces a precautionary element by setting strict substantive standards (see below), the Directive predominantly follows a reactive, 'end-of-the-pipe' approach. The water companies and the consumers have to bear the costs of cleaning polluted water, which is in contrast to the precautionary the 'polluter-pays' principle, which aims at preventing water pollution in the first place. With regard to the policy instruments, the Directive contains some elements of procedural regulation. Annex II and III prescribe how often and by what means the monitoring of the water quality is to be carried out and it gives reference to a method of analysis for each parameter. Member states must monitor compliance with the conditions of authorization and the effects of discharges on groundwater, keep an inventory of authorizations, and supply the European Commission with any relevant information at its request.

---

[19]   80/778/EEC; OJ L 129, 30.8.80.

[20]   The other two Directives are the Directive on the Quality of Surface Water 75/440 (OJ L 194, 15.7.75) and the Directive on Medicinal Waters and Mineral Waters 80/777 (OJ L 299/1, 30.8.80).

[21]   The water intended for direct human consumption is subject to all mandatory values of maximum admissible concentration in the Directive. For the water used in food production, only the toxic and microbiological parameters are mandatory. The member states are free to set their own values for the organoleptic and physiochemical parameters as well as for undesirable substances.

The Drinking Water Directive was a response to the mounting concern about the increased reuse of waste water for potable supply and the rising number of new organic and other trace substances entering into the water supply. The 64 water quality standards are drawn from the World Health Organization (WHO) 1970 Drinking Water Standards. The negotiations for the Directive were lengthy and cumbersome. The member states disagreed on a variety of parameters for both cost and public health considerations. The Netherlands pushed for stringent standards to fight the pollution of the river Rhine but was not able to overcome resistance from other countries, who wished to focus on the public health dimension of the policy. The UK campaigned for a more lenient standard on lead, since the proposed EU standard was half of what the WHO recommended as a maximum allowable concentration (25 as opposed to 50 μ/l). While the British government did not succeed, the Directive contains a number of provisions to facilitate compliance with the more stringent EU standard (Haigh 2001: 4.4-5). The UK, together with Italy, also rejected the allowable concentration for pesticides, which did not coincide with WHO guidelines either. The British and the Italians argued that there was no scientific evidence to justify a general prohibition on levels of pesticides wherever they could be detected. They demanded more realistic standards, which would keep compliance costs at a reasonable level. Germany and the Netherlands, by contrast, took the position that as a matter of principle, drinking water should be as close to its natural state as possible and therefore not contain any pesticide residues (Weale et al. 2000: 358). The precautionary arguments put forward by the Germans and the Dutch carried the day. The Directive largely followed the model of the German Drinking Water Ordinance passed in 1980 (see below). Compromise among the member states was facilitated by the technological and scientific uncertainty regarding some of the quality standards. In some cases, the national governments were simply not able to anticipate the compliance costs (Haigh 2001: 4.4-5).

In 1998, the Drinking Water Directive was revised in light of new scientific and technical evidence.[22] The new Directive will replace its predecessor as from 2003. It reduces the number of parameters down to 53, of which nine are newly introduced. It also distinguishes between substances with mandatory and indicator limits. Finally, some parameters, like lead, become more stringent (cf. Haigh 2001: 4.4-6).

Spain: Neither Pull nor Push

In Spain, the quality of drinking water became the subject of comprehensive regulation in 1982. Four years before Spain's accession to the EC, the Technical Health Regulation of the Supply and Quality Control of Drinking Water[23] was

---

[22] 98/83/EC, Council Directive of 3 November 1998 on the quality of water intended for human consumption, OJ L 330, 05.12.1998.

[23] *Reglamentación Técnico-sanitaria para el abastecimiento y control de calidad de las aguas potables de consumo público*, RD 1423/1982, 18.6.1982; BOE N° 151, 29.6.1982.

enacted. The regulation systematically introduced a list of binding standards for drinking water quality and set up procedures for monitoring compliance. It is striking how closely the regulation followed the Drinking Water Directive, both in structure and content. [24] The definition of '*aguas potables de consumo público*' (Art. 2.1.), for example, was identical with the definition of 'water intended for human consumption' in the Directive (Art. 2). Like the Directive, the regulation contained a list of organoleptic and physico-chemical parameters, undesirable and toxic substances, and microbiological parameters. For each parameter, both a guide level (*orientadores de calidad*) and a maximum admissible concentration (*caracterers tolerables de calidad*) were specified. The Spanish regulation, however, failed to set a number of mandatory values specified in the Directive, including silica, sodium, potassium, and total organic carbon. The maximum admissible concentration allowed for some parameters was considerably higher than in the Directive (e.g. for phenols, chlorides and sulphates). The regulation followed the method of analysis of the Directive, making the distinction between initial analysis, minimum, current and periodic monitoring and prescribing patterns and frequency of standard analyses. The lists of parameters to be monitored and the minimum frequency of standard analyses largely coincided, although the Directive was in some cases more demanding. The rules for granting temporary transgression of the maximum admissible concentration of the Spanish regulation and the Directive were identical.

All in all, the Spanish drinking water regulation was clearly oriented towards the European Directive. It anticipated many of the European requirements adopting its problem-solving approach and policy instruments, which by and large coincided with Spanish regulatory practices. As a result, the policy misfit between Spanish and European drinking water regulations was relatively low and did not seem to produce much pressure for adaptation. The Spanish administration could easily absorb the European drinking water policy into its regulatory structures. It only had to adopt some of the limit values and introduce a number of new ones, respectively. A circular of 1987[25] brought the Spanish regulation into formal compliance with the Drinking Water Directive. Three years later, the Directive was officially transposed into Spanish legislation.[26] The transposition act simply merged parts of the Spanish regulation and the European Directive by taking them over literally. Next to the legal adaptation of guiding values, the transposition legislation gave rise to two minor administrative changes. First, the Departments of Public Health of those Autonomous Communities that took on the responsibility for monitoring water quality must report on the exceeding of European drinking water standards to the national Ministry of Public Health and Consumption, which transfers the data

---

[24]   This was not the first case where Spain adjusted its water legislation to European standards. In 1981, the Directive on Medicinal Waters and Mineral Waters of 1980 had been transposed into Spanish law (RD 2119/1981, 24.7.1981).

[25]   *Orden de 1 de julio de 1097*, BOE N° 163, 9.7.1987.

[26]   *Real Decreto* 1138/1990, BOE N° 226, 20.9.1990.

to the European Commission (Art 30 RD 1138/1990). Second, the regional Departments of Public Health can authorize temporal derogations from maximal admissible concentrations, which had not been possible before (Art. 3.2. RD 1138/1990).

Since the legal and administrative structure for regulating and monitoring the quality of drinking water had already been in place for some years before the Directive was finally transposed, formal implementation did not cause significant costs, neither material nor political. Unlike in Germany, Spanish policy-makers did not face much opposition in bringing national standards up to the more stringent European levels. Spanish water suppliers were hardly aware of the compliance costs, which practical application and enforcement of the new European standards would incur on them. Moreover, the lax monitoring and enforcement practices of public authorities at the local level made it seem unlikely that they would have to fully comply.[27]

Indeed, insufficient monitoring capacities have rendered it difficult to assess Spain's compliance with the Drinking Water Directive. Reliable data on water quality have hardly been available. Monitoring capacities suffer from an uneven geographical distribution of measurement stations and a measurement technology which is often not up to standard (OECD 1997; Instituto para la Política Ambiental Europea 1997). The measurement of some parameters set by the Drinking Water Directive, for instance, requires sophisticated technological equipment that is not only expensive but also presupposes manpower and expertise which water companies and municipalities often do not possess. Both the Commission and the Spanish authorities have been aware of the problem. In 1994, the Spanish Directorate General of Environment published a report on the Spanish monitoring networks for water quality, which found considerable deficiencies with respect to the technical standard of measurement techniques and territorial coverage. To remedy the situation, an investment of EUR 60.1 million was envisioned to extend and upgrade the national networks (*Medio Ambiente en España 1994*: 146-148). At the local level, problems of insufficient monitoring capacities are even more pronounced. The municipalities have neither sufficient technical equipment and qualified staff nor the financial means to acquire them (OECD 1997). The Autonomous Communities have not always been willing to take over monitoring responsibilities from the local authorities. But even if they do, they often use different measurement methods, which render the comparison of regional air quality data difficult and makes compliance assessments for Spain as a whole nearly impossible.

While it appears unlikely that Spain has effectively applied and enforced the Drinking Water Directive, the Spanish government only began to feel some external pressure for adaptation 12 years after it had joined the EC. In 1998 and 1999, respectively, the Commission opened two infringement proceedings for the improper application of the Directive. Both cases resulted in a Reasoned Opinion, which the Spanish government received in July 2000. The cases are still open at

---

[27] Interview with the Catalan Ministry of Public Health, Barcelona, 03/97.

the time of writing. The long absence of pressure from above is explained first, by the *formal* compliance of Spanish drinking water legislation with European regulations and the reluctance of the Commission to interfere with practical application. Second, there has been little domestic mobilization that would have induced the Commission to take action. Spain has been mostly concerned with problems of water quantity rather than water quality. As consumption has risen with expanding urban populations and enlarged agricultural areas under irrigation, water has become increasingly scarce. Growing competition for water has provoked serious conflicts between users and their regional governments. Some Autonomous Communities have started war-like confrontations over the distribution of water (*guerras del agua*, cf. Grau i Creus 2001). Problem pressure is still too low for environmentalists and citizens to mobilize against the ineffective monitoring and enforcement of drinking water standards.[28]

Germany: Push without Pull

German law did not contain any general provisions on the quality of drinking water. Different regulations are found in sectoral laws, of which the Federal Law on Contagious Disease (*Bundesseuchengesetz*) is the most important. Its executing regulation, the Drinking Water Ordinance (*Trinkwasserverordnung*) of 1980, set some specific requirements for drinking water.[29] The Ordinance reflected the legalistic and interventionist policy approach typical for German environmental policy-making. It stipulated that drinking water should not contain substances detrimental to human health. Most importantly, drinking water had to be free of pathogenic agents. Accordingly, the Ordinance defined 16 uniform and legally binding standards for the permissible concentration of substances in drinking water. Next to microbiological parameters, they included limit values for chemical and radioactive substances. Moreover, the Ordinance prescribed the substances that could be used for water treatment and defined the procedures and technologies to be used to monitor the water quality. Water providers were obliged to comply with these standards and use the best available technology where required by the Ordinance. They had to regularly control the water quality and report to the health authorities, which were in charge of monitoring the water quality at the places of consumption.

In principle, the Drinking Water Directive corresponds to the German regulatory approach. Nevertheless, some of the European standards as well as certain procedural requirements produced substantial policy misfits requiring modifications of the German Ordinance that entailed considerable costs for both public and private actors. First, unlike the German Drinking Water Ordinance, the European

---

[28] Interviews with water experts of ADENA-WWF, AEDENAT, and CODA, Madrid, 03/97.

[29] *Verordnung über Trinkwasser und über Brauchwasser für Lebensmittelbetriebe vom 31. Januar 1975, abgeändert am 25. Juni 1980 und 22. Mai 1986, BGBl. I: 76.*

quality standards not only account for public health requirements but also aim at protecting the environment. Drinking water sources must be sufficiently free from contamination to allow for appropriate water treatment. The German Drinking Water Ordinance, by contrast, was exclusively oriented towards public health concerns. As a result, it did not regulate those European parameters that were included in the Directive for environmental reasons. The monitoring and enforcement of altogether 64 European parameters – compared to the 16 in the German Ordinance – required public authorities to spend several billion D-Marks on additional staff-power and measurement technology (Kromarek 1987: 41). So did water suppliers in order to execute the analysis of a whole variety of new parameters. Second, the European limit values for pesticides and a few other parameters were more stringent. The limit value for nitrate, for example, was 80 per cent more demanding in the European Directive than in the German Ordinance ($50\mu g/l$ as opposed to $90\mu g/l$), and for mercury even four times ($1\mu g/l$ as opposed to $4 \mu g/l$). In order to meet these more stringent standards, industry had to invest in new abatement technologies, while farmers had to restrict the use of fertilizer, the major source of nitrate emissions. According to some estimates, the immediate enforcement of the European limit values for pesticides and nitrates would have required the closure of many wells, particularly in scarcely populated areas, where people were not connected to the public water network (Müller-Brandeck-Bocquet 1996: 143).

It took Germany more than six years and the opening of an infringement proceeding by the Commission in 1987 to revise its legislation and bring it in line with European obligations. The Directive was finally transposed in 1986, four years after the deadline had expired.[30] But the revision of the Drinking Water Ordinance included only 37 of the 64 European parameters. The omission was deliberate, since Germany intended to meet only the public health requirements of the Directive (Kromarek 1987: 43). Consequently, four European parameters concerning toxic substances were not part of Annex II listing the chemical substances in the German Ordinance. They were replaced by parameters listed in the Directive as substances undesirable in excessive amounts (nitrate, nitrite, fluoride, and organ chlorine). While the limit values for some toxic parameters were more stringent than in the Directive (cadmium, lead, pesticides, PCB), the Ordinance only adopted two of the six microbiological parameters of the Directive that were considered relevant to public health ignoring, for instance, the minimum required concentration for the total hardness of water.[31] The Ordinance took over the physico-chemical parameters of the Directive (with two exceptions) and adopted the corresponding limit values. But the *Länder* governments were entitled to change these limit values if required by regional circumstances and if these exemptions did not constitute a risk for public health (Annex II). Not only did the Ordinance define ex-

---

[30] *Verordnung über Trinkwasser und über Wasser für Lebensmittelbetriebe vom 22. Mai 1986,* BGBl. I: 76.

[31] For reasons of public health, the Ordinance introduced two parameters not included in the Directive: chlorine and chlorine dioxide.

emptions more broadly than the Directive, which restricts exemptions to cases of emergencies defined as unanticipated incidences of short-term nature (e.g. thunderstorms). Unlike required by the Directive, the Ordinance did not oblige the *Länder* to inform either the federal government or the European Commission about granted exemptions. Moreover, *Bund* and the *Länder* agreed to enforce the pesticide standards only from 1989 on in order to give agriculture some time for adjustment. Monitoring regulations followed European requirements but did not fully conform to them. The Ordinance was more demanding than the Directive since it prescribed regular minimum controls for chemical parameters once or twice a year while the Directive only requires occasional controls. With regard to the monitoring of the other parameters, however, German regulations were less stringent because they allowed for a reduction of controls irrespective of whether maximum admissible concentrations were complied with (cf. Kromarek 1987: 43-44).

To conclude, while the German Drinking Ordinance partly complied with the Directive, and in some instances entailed even more stringent regulations, it contained serious omissions in order to avoid costly adaptation, particularly in the agricultural sector. German authorities justified the delayed and incomplete transposition of the Drinking Water Directive with the high quality of drinking water, which Germany had enjoyed since more then 100 years. They rejected some parameters as irrelevant to public health refusing to acknowledge the environmental considerations of the Directive. For the omission of others, German authorities argued that their maximum admissible concentration was too small to be detected by existing measurement technologies. Germany wanted to wait with the transposition of the Directive until adequate measurement technologies would be available. Insufficient monitoring capacity would have 'delegitimized the German regulatory practices by questioning the credibility of law' (Knill and Lenschow 2001: 130). At the same time, German authorities criticized the measurement procedures imposed by the Directive as outdated and no longer up to the standards of the best available technology – after the long time that the negotiations on the Directive had taken (Knill 2001: 154-155). Finally, German authorities complained that the Directive was not sufficiently integrated with other European water policies and would not leave the member states enough time to adapt to the stringent standards (Kromarek 1987: 46).

Despite several violations against the Drinking Water Directive, domestic actors largely refrained from pressuring German policy-makers and administrators to effectively implement and apply European obligations. Unlike in Spain, Germans have been concerned about the quality of drinking water. The regular and frequent testing of water quality with modern measurement technology as well as the obligation to report any breaches to the public health authorities, which in turn have to notify the public in case of potential danger, have led to an almost immediate exposure of even temporary violations of the Directive. The public can be easily informed and mobilized on the issue. At the same time, there is the danger of people overreacting, which might impair a reasonable compromise between health concerns of the consumers, on the one hand, and the economic concerns of water suppliers and the polluters, on the other (Kromarek 1987: 46). German environmental groups have largely supported the gradual adaptation to European standards favoured by the public administration. They considered the cooperation between public authorities, water suppliers, and polluters as more effective to facilitate the necessary adaptations than the immediate enforcement of European obligations.[32]

While domestic pressure from below has been largely absent, the Commission pushed from above. In 1987, it opened proceedings for the incomplete transposition of the Directive. The case was referred to the European Court of Justice in 1990. Facing a conviction, Germany finally abandoned its resistance. The second revision of the Drinking Water Ordinance included all the parameters and conformed to the measurement requirements of the Directive.[33] It also brought the exemption clause in correspondence with European prescriptions. The Commission conceded that with the revision of the Drinking Water Ordinance of 1990 Germany formally complied with the Directive. It claimed, however, that Germany had delayed implementation and would not correctly apply the Directive either (see below). The ECJ upheld the complaints of the Commission in its 1992 judgement.[34]

Administrative practice has not changed much after Germany had formally complied with the Directive. The practical application and enforcement of certain limit values have remained a problem (Länderarbeitsgemeinschaft Wasser 1997). This is particularly true for nitrate. German water suppliers already had problems in meeting the far more lenient standards of the German Drinking Water Ordinance. The concentration of nitrate in German drinking water has further increased in areas with intensive farming, dairy, and wine production. Water pollution from diffuse agricultural sources is one of the major environmental challenges for Germany, particularly the growing nitrate pollution of groundwater resources representing more than 70 per cent of German drinking water supply. In a 1995 national survey, about ten per cent of measurements exceeded the European drinking water limit value for nitrate (50 μ/l) and pesticides (0.1 μ/l, OECD 2001: 62).

Both water providers and environmental organizations pushed for stricter emission standards for industrial polluters and compulsory limits on the use of fertilizers for farmers. The demands caused strong resistance from the chemical industry. And the farmers were only willing to change their production practices in exchange for financial compensation. The fifth amendment of the German Water Resources Management Act (*Wasserhaushaltsgesetz/WHG*) allowed compensation payments to farmers. While this appeared to be against the 'polluter-pays-principle', the German government was not willing to provoke a major conflict with the federal and the regional ministries of agriculture and their powerful clientele, the farmers associations (Rüdiger and Krämer 1994: 69).

---

[32] Interviews with water experts of WWF, Bremen 01/98; DNR and BUND, Bonn 10/97.

[33] *Neufassung der Verordnung über Trinkwasser für Lebensmittelbetriebe vom 12.12.1990*, BGBl. I: 2613-2629 with a correction of 23.1.1991, BGBl. I: 227.

[34] C-1990/237, 24.11.1992.

Instead of imposing stricter emission standards, some *Länder* introduced compulsory consumption fees (*Wasserpfennig, Wasserentnahmeengelte*) to help finance investments in new abatement technologies and compensation payments for farmers willing to reduce the use of pesticides and fertilizers. Other *Länder* have sought to ease the burden for the local water supplying companies by mediating cooperative agreements between local water providers and potential polluters such as farmers. For instance, Bavaria has more than 2700 local water supplying companies (mostly public) of which about 25 per cent had problems in complying with the European limit values for pesticide (Schretzenmayr 1989). In order to improve compliance, the Bavarian government developed so called nitrate programmes, voluntary agreements between farmers and water companies in which the water companies would provide financial compensation and expertise to farmers as to help them reduce the emission of nitrates. Similar arrangements are found in other *Länder* (Lenschow 1997). Moreover, regional and local authorities have often negotiated temporary suspension of certain standards with the water suppliers adopting a broad interpretation of what constitutes an 'emergency' defined by the Drinking Water Directive as the precondition for granting any exemptions (Gran 1989).

In its 1992 judgement, the ECJ upheld the view of the Commission, which considered such exemptions a serious infringement of the Directive. In light of another infringement proceeding, which the Commission had opened for not properly applying the Directive and which resulted in a Reasoned Opinion in 1993, Germany officially gave in to the view of the Commission. Yet, the *Länder* administrations have kept insisting on the necessity of exemptions – negotiated on a case by case basis between the *Länder* administration and the water companies – to ensure an effective implementation of the Directive.[35] Regional authorities would only grant exemptions if they did not cause a danger for public health and were linked to corrective measures to be negotiated with the water supply companies. Such a flexible approach would allow the gradual compliance with the Directive rather than simply closing down water supply locations. The *Länder* have also defied the formalistic approach of the ECJ and the Commission to granting temporary exemptions as an 'assault' against German administrative practice, which would result in a paradoxical situation of favouring environmental laggards, like Greece and Italy, where European Directives are simply copied into national law but not enforced, over environmental leaders, like Germany, which effectively comply with European regulations (Börzel 2002b: 159). As yet, the Commission seems to tolerate the German practice of granting 'flexible' exemptions. It has so far refrained from opening infringement proceedings. Germany, in turn, timely transposed the revision of the Drinking Water Directive.[36]

---

[35] Interviews in the Ministries of Environment and of Public Health of Bavaria, Munich 10/97, North-Rhine-Westphalia, Düsseldorf 09/98, and Baden-Württemberg, Stuttgart 03/98.

[36] *Verordnung zur Novellierung der Trinkwasseverordnungr vom 21. Mai 2001*, BGBl. I: 959-980.

In sum, the Drinking Water Directive caused misfits for Spain and Germany alike, resulting in serious problems of compliance for both countries. Spain, however, has not faced any significant pressure for adaptation, neither from below nor from above, to face the compliance costs. Spanish environmental groups do not perceive water quality as a major problem. In the absence of significant domestic mobilization, the Commission has largely refrained from denouncing Spain for infringements of the Directive. Germany did not face much domestic pressure for adaptation either. Environmentalists largely supported the gradual adaptation to European standards. But in contrast to Spain, the Commission exerted considerable pressure from above for the incomplete transposition of the Directive. This 'push without pull' ultimately brought Germany in formal compliance with the Drinking Water Directive. In the absence of domestic mobilization, however, full compliance in practical application and enforcement remains questionable.

Air Pollution Control: Spanish Misfit and German Fit

The Policies and their Developments

There are two Directives which have been of particular importance for the EU policy on the combating of air pollution: the Directive on the Limitation of Emissions of Certain Pollutants into the Air from Large Combustion Plants and its 'Mother Directive', the Directive on the Combating of Air Pollution from Industrial Plants.

The *Directive on the Combating of Air Pollution from Industrial Plants* (Industrial Plant Directive), which the Council adopted on 28 June 1984,[37] provides a framework legislation for preventing and reducing air pollution caused from industrial plants (energy industry, production and processing of metals, manufacture of non-metallic mineral products, chemical industry, waste disposal). It is the first European response to the problem of acid deposition and the death of forests.

The Directive does not set any substantive emission standards but establishes an administrative framework containing procedural requirements for the authorization of the operation of any new industrial plant and the substantial modification of existing plants. It thereby follows the principle of procedural regulation embracing a precautionary and technology-based problem-solving approach. From 1 July 1987, the authorization of a new plant, or the substantial modification of an already existing plant, especially if they fall in one of the 19 categories specified in Annex I of the Directive, must only be granted if 1) all appropriate preventive measures against air pollution were taken, including the application of the best available technology not entailing excessive costs (BATNEEC), 2) the emission of the plant will not cause any significant air pollution, especially with regard to those polluting substances listed in Annex II of the Directive, and 3) none of the emission or

---

[37]   84/360/EEC; OJ L 188, 16.7.84.

air quality limit values applicable will be exceeded. The member states are to ensure that applications for authorization and the respective decision of the competent authority are made available to the public concerned. The Directive establishes the categories of industrial plants that require prior authorization. It also makes specific provisions for existing industrial plants predating 1 July 1997. The member states must implement measures in order to gradually adapt existing plants falling into the categories specified in the Directive to the best available technology. The adaptation of existing plants is to be carried out by taking into account the technical characteristics of the plant and the nature and volume of its polluting emissions. Adaptation, however, must not impose excessive costs on the plant concerned. The Directive finally stipulates the possibility for the Council to adopt 'daughter directives' fixing 'emission limit values', based on the best available technology not entailing excessive costs, but only if it appears to be strictly necessary and by the unanimous vote of the member states.

The Industrial Plant Directive is replaced by the new Directive on Integrated Pollution Prevention and Control (IPPC),[38] which involves an integrated approach to the permitting of industrial sites. It is intended to change and harmonize the environmental regulation of industry in the member states by introducing common requirements for issuing permits to large sources of industrial pollution. Emission limits to all media are to be considered simultaneously. The IPPC Directive already supersedes the authorization procedure of the Industrial Plant Directive for 'new' plants, while 'existing' plants will be brought under the new regime by the end of October 2007.

The first time the Council set emission limit values was four years later when it adopted the *Directive on the Limitation of Emissions of Certain Pollutants into the Air from Large Combustion Plants* (Large Combustion Plant Directive).[39] The Large Combustion Plant Directive seeks to limit acidifying emissions, especially sulphur dioxide, nitrogen oxides, and particulate matter from fossil-fuelled power stations. It shares the precautionary, technology- and emission based problem-solving approach of its Mother Directive. Combustion Plants with a rated thermal output of 50 MW or more have to apply the best available technology not entailing excessive costs (BATNEEC) to reduce their emissions. Measuring methods used for monitoring compliance also have to conform to BATNEEC. Although the Directive contains procedural requirements for the monitoring of plants (Annex IX), it predominantly relies on command-and-control substantial regulation in order to prevent and reduce air pollution. For 'new' large combustion plants with a thermal output greater than 50 MW, Annexes III to VII of the Directive define binding emission limit values for three pollutants: sulphur dioxide (SO_2), nitrogen oxides (NO_x), and dust. New plants have to comply with these emission limit values and use BATNEEC to reduce air pollution. 'Existing' plants for which the original construction licence, or the operating licence, was granted before 1 July 1987 are

---

[38]   96/61/EC: OJ L 257, 10.10.1996.

[39]   88/609/EEC; OJ L 336, 7.12.88.

not subject to emission standards as such, but the member states have to progressively reduce their total annual emissions. By 1 July 1990, the member states should introduce national emission reduction plans setting out the timetable and the implementing procedures for the total emission reduction of the relevant gases. By applying the best available technology, the emission for SO_2 and NO_x have to be reduced by an increasing percentage over 1980 levels in three five year intervals – from 1988 for 1993, 1998, and 2003. The emission ceilings (national bubbles) laid down in Annex I and II of the Directive vary among the member states according to their economic, energy and environmental situation. For example, Belgium, Germany, France, and the Netherlands have been assigned targets of a 70 per cent reduction in SO_2 emissions between 1980 and 2003. Denmark, Italy, Luxembourg and the UK each have to reduce their emission by at least 60 per cent over the same period. Greece, Italy, and Portugal, by contrast, are allowed some increases. Although setting uniform and legally binding standards, the member states dispose of some flexibility in reducing the emissions of existing plants. Within their 'national bubbles', they are free to vary the stringency of emissions standards applied to any particular plant. Yet, since energy generation has depended on high sulphur coal in most member states,[40] they could only meet the reduction targets within the time-frame set by installing flue-gas desulphurization (Boehmer-Christiansen and Skea 1991).

Finally, the member states must inform the European Commission on their programmes for total emissions reductions by 31 December 1990. One year after the end of each phase, the Commission must be sent summary reports on the implementation of the programmes. On the basis of these reports, the Commission proposed a revision of the Directive. The new Directive was passed in October 2001.[41] It updates the national emission ceilings as well as emission limit values applicable to plants licensed after November 2002 and extends the scope of Directive 88/609 to include gas turbines from the same date. It also sets more stringent requirements for plants for which a license was issued prior to 1 July 1987 (existing plants).

The new Directive on National Emissions Ceilings, finally, requires the member states to limit their annual national emissions of the pollutants sulphur dioxide (SO2), nitrogen oxides (NOx), volatile organic compounds (VOC) and ammonia (NH₃), including contributions from large combustion plants, to amounts not greater than the emission ceilings laid down in Annex I of the Directive by the year 2010 at the latest.[42]

The two Air Pollution Control Directives are to tackle one of the principal causes of acid rain. In the 1970s, Scandinavia, Germany, and other European countries faced significant acid rain damage of their freshwater lakes, streams, forests, and historical buildings. Scientists and policy-makers framed the environ-

---

[40] France and Belgium were the exceptions since they had large nuclear programmes.

[41] 2001/80/EC, OJ L 309, 27.11.2001.

[42] 2001/81/EC, OJ L 309, 27.11.2001.

mental problem as transnational, pointing to the long-range transport of certain air emissions, such as sulphur dioxides (SO_2) and nitrogen oxides (NO_x). Particular blame was placed on the emissions of fossil-fuelled power stations and other large combustion plants (e.g. oil refineries) operated by heavy industry and power utilities (Zito 2000: 49-52). The UK and Germany, as two major pollution emitters, had initially rejected a link between their industrial activities and the acidification damage suffered by their regional neighbours. The forest death alarm (*Waldsterben*) in the early 1980s, however, made Germany reconsider its position and announce a radical policy change at the Stockholm Conference on the Acidification of the Environment. In 1982, it revised and modernized its air pollution legislation with the adoption of the Large Combustion Plant Ordinance. In order to avoid competitive disadvantages for its industry and minimize costs of compliance with a European solution alien to its own regulatory style, the German government sought to impose its domestic solution for combating industrial emissions on the other member states. In June 1982, Germany submitted a memorandum to the Council advocating effective policies against air pollution and calling on the Commission to submit a proposal for a basic Directive before the end of 1982 (Liefferink 1996: 117). The Commission responded with the proposal of the Industrial Plant Directive, which it presented in April 1983. The Netherlands and Denmark, which have a similar regulatory tradition as Germany, supported the German attempt to upload its policies to the European level. Even the UK, which had earned itself the reputation as the 'dirty man of Europe' at the time, accepted the Industrial Plant Directive whose procedural regulations appeared by and large in line with British practice. Moreover, the Directive paid tribute to the British approach of weighing the economic costs of protection measures against their environmental benefits, replacing the German concept of 'the state of the art' by 'the best available technology not entailing excessive costs' (BATNEEC). Also, the fixing of substantial emission limit values would require unanimity in the Council taking into account the nature, quantities and harmfulness of emissions as well as considerations of BATNEEC. Finally the implementation deadline was postponed by three years to 30 June 1987 (Haigh 2001: 6.9-3).

But the British government fiercely opposed the proposal for the Large Combustion Plant, which was largely modelled after the German Large Combustion Plant Ordinance (Zito 2000: 58-61). Imposing stringent emission standards in order to prevent air pollution conflicted with the traditional British approach guided by ambient air quality, strict scientific causality, and economic proportionality. In contrast to the German approach, the British premise is that the environment can absorb a certain emission load without suffering harm. This justifies that emissions are not avoided at any costs; rather the aim is to achieve a low-cost intensive use of the environment, which can differ depending on local conditions, technology costs, and the economic situation of the firm concerned (cf. Héritier 1996: 77-87). Thus, the British government and its energy industry argued that the large compliance costs anticipated by heavy investments in abatement technology were not justified given the absence of sufficient scientific evidence that large cuts in sulphur dioxide emissions would help to solve the problem of forest death as claimed by the

Germans (Boehmer-Christiansen and Skea 1991: 206-210). The UK was not only the largest sulphur dioxide emitter. Large combustion plants accounted for over 80 per cent of British SO_2 emissions, compared to less than 70 per cent in Germany and the Netherlands (Weale et al. 2000: 392). Moreover, electricity generation industry was nationalized as a result of which investment in abatement technology would show up as expenditures in public accounts. Spain joined the British opposition as soon as it had entered the European Community in 1986. Like the other industrially less developed member states (Greece, Ireland and Portugal), Spain emphasized its rapidly growing energy needs, which it predominantly met by coal-fired power stations. After five years of tough negotiations and fierce conflict, the Council of Ministers finally adopted the Large Combustion Plant in 1988. Spain's consent was literally bought by financial compensations and a temporary derogation from the full application of the SO_2 limits (cf. Héritier 1996: 184-203; Zito 2000: 58-68). Being isolated in its opposition of the Directive, the British government finally gave in not without having negotiated some concessions to reduce its compliance costs. As the largest SO_2 emitter it was able to secure for itself smaller percentage reductions than other heavy polluters, such as Germany. Moreover, it achieved derogation for new plants burning indigenous high sulphur coal. The UK had also struck a package deal with Germany linking British approval of the LCP Directive to the German agreement on looser emission standards for small cars (Zito 2000: 64).[43] The consent of the three smaller cohesion countries, Greece, Portugal, and Ireland, was won by allowing them to increase emissions as to not inhibit their economic development.

Spain: Complete Misfit

The Industrial Plant Directive: neither pull nor push Triggered by the developments in international environmental policy and a concern about an increasing level of CO_2 emissions, Spain initiated a series of laws to combat air pollution in the early 1970s.[44] The Air Protection Act[45] of 1972, which provides a first legal

---

[43] Directive 89/458 amended EC legislation on emissions from vehicles by introducing mandatory emission limits for cars whose engine size was less than 1.4 litres, in line with the then US and Japanese standards, effectively requiring the use of controlled three-way catalytic converters (Haigh 2001: 6.8-5).

[44] Spain enacted the first national regulation on fighting air pollution already in 1961 by a Decree, which approved the Regulation on Irritating, Unhealthy, Harmful and Dangerous Activities (*Reglamento de Actividades Molestas, Insalubres, Nocivas y Peligrosas, Decreto de 24 de 4 abril de 1961*, 30.11.1961, BOE N° 292, 7.12.1961). Art. 13 of this regulation requires that any activity considered unhealthy due to its production of atmospheric pollutants should be subject to public control (cf. the section on EIA).

[45] *Ley 38/1972, de 22 de diciembre de Protección del Ambiente Atmosférico*, BOE N° 309, 16.12.1972, which is further specified by a ministerial circular of October 1976 (*Orden de 18 de octubre de 1976*, BOE N° 290, 3.12.1976).

framework for combating air pollution, lays out some general regulations for the authorization of new industrial installations and the modification of existing plants.

A Decree of 1975[46] regulated in more detail the authorization of industrial installations. Like in the procedure prescribed by the Industrial Plant Directive of 1984, all new installations and the modification of existing industrial plants that were likely to increase their emission level listed in Annex II of RD 833/1975, were subject to prior authorization by the municipalities. In most cases, the local license required a special authorization for activities potentially polluting the atmosphere, according to the Regulation on Irritating, Unhealthy, Harmful and Dangerous Activities. Depending on the type of activity, this authorization had to be given by the national or regional Ministries of Industry. Authorization should only be granted to industrial installations that did not exceed the emission limits and air quality standards specified in the Annex I of Titulo II and Annex IV of Titulo V of the Decree, which covered more or less the most important polluting substances listed in Annex II of the Industrial Plant Directive. Moreover, plants had to adopt certain measures and technologies allowing for the dispersion of polluting substances as to prevent the exceeding of air quality standards (height of chimney, temperature and speed of emission expulsion). Besides, the environmental impact of an industrial plant had to be assessed by the national or regional Ministry of Industry, in cooperation with the environmental authorities. This 'environmental impact assessment' *(evaluación del impacto ambiental)*,[47] also included the obligation to consider corrective measures which could be made conditional for authorization. Finally, the Decree 833/1975 prescribed the establishment of a national network of fixed and mobile monitoring stations *(Red Nacional de Vigilancia y Prevensión de la Contaminación Atmosférica)*. The monitoring network was placed under the supervision of the environmental authorities at the regional level.

At first sight, Spanish legislation on combating air pollution appears to fulfil the major requirements of the Industrial Plant Directive (Alonso García 1994). A closer look at the regulations, however, reveals important deficiencies. Spanish legislation did not share the precautionary approach of the Directive. First, the formal requirements of the impact assessment as well as the consideration of corrective measures were certainly not sufficient to meet the precautionary regulations of the Directive (Art. 4.1 and 4.2.). Likewise, dispersion measures did not really qualify as appropriate preventive measures against air pollution. Nor did the designation of 'Polluted Air Zones', in which air quality standards were exceeded for a certain number of days, and the declaration of 'Emergency Situations' in cases where air quality standards were considerably exceeded. Second, Spanish legislation acknowledged the importance of technological progress in combating air pollution. But it linked any reference to available technology to economic imperatives

---

[46]   *Decreto 833/1975 de 6 de febrero*, BOE N° 96, 22.4.1975.

[47]   *Orden de 18 de octubre de 1976*, BOE N° 290, 3.12.1976.

against which environmental protection had to be compromised.[48] Moreover, the criterion of 'best available technology' (BAT) was nowhere defined nor was its application explicitly required.

Despite these considerable misfits between the precautionary problem solving approach of the EU legislation and the reactive Spanish regulations, the Industrial Plant Directive has never been transposed into Spanish law. In 1987, the Director- ate General of Environment had announced the intention to amend existing legis- lation in order to comply with the Directive acknowledging the incompatibilities between Spanish and European regulations (Bennett 1991: 184). The Ministry of Industry, however, which has been in charge of compliance with the Directive, contented that Spanish legislation already fulfilled European requirements.

Since the Industrial Plant Directive has never been transposed, those parts, which do not correspond to Spanish legislation, have not been practically applied either. In order to avoid additional working load and escape opposition of industry, public administration has largely refrained from systematically considering preventive measures in the authorization of industrial plants. Nor has it enforced the use of BAT on Spanish industrial plants. Some environmental administrators defy the BAT concept as a 'Northern European invention', which does not fit Spanish environmental policy.[49] Moreover, the consequent enforcement of BAT would require the state to provide ample subsidies since Spanish industry is predominantly made up by small and middle-sized enterprises, which often do not have sufficient financial resources to invest in new abatement technology.[50]

Environmental groups have been aware of the ineffective implementation of the Industrial Plant Directive. But they refrained from mobilizing against public ad- ministration because they wanted to concentrate their resources on more 'important issues'.[51] Air pollution has not been a high-ranking issue neither in Spanish public opinion nor on the political agenda of the environmentalists. The Commission, in turn, has apparently accepted Spain's claim that existing legislation already com- plied with the requirements of the Directive. It has not opened infringement

---

[48] See for instance the preamble of the *Ley de 22 de diciembre de Protección del Ambiente Atmosferico*, BOE N° 309, 26.12.1972; cf. the preamble, Titulo I, B, and Art. 48.3. of the *Decreto 833/1975 de 6 de febrero*, BOE N° 96, 27.4.1975.

[49] In responding to a written request for information on the implementation of BAT in Catalonia, I received three letters by the Catalan Ministry of Environment: two inform- ing me that there are no specific regulations on the application of BAT, and the third ex- plaining to me that the BAT concept starts from '*realidades y mentalidades germanicas muy poco adaptables a la realidad de las problematicas del sur de Europa*' (Germanic realities and mentalities which are hardly applicable to the real problems of Southern Europe).

[50] Interviews with the air pollution control unit of the Catalan Ministry of Environment, Barcelona, 03/97, and, representatives of the Spanish Business Association (CEOE), Madrid, 03/97.

[51] Interview with an air pollution expert of CODA, Madrid, 03/97.

proceedings for faulty transposition of the Directive.

In sum, there is a considerable misfit between the Industrial Plant Directive and Spanish air pollution control legislation. Public administration has refrained from effective implementation due to the adaptational costs. In the absence of pressure from above by the Commission for incomplete transposition, and pressure from below by societal actors for non-application and non-enforcement, Spain's compliance with the Industrial Plant Directive has remained low.

The Large Combustion Plant Directive: pull without push　Spanish legislation entailed both quality and emission standards in order to control air pollution. The already mentioned Decree 833/1975 laid down a number of air quality standards, such as sulphur dioxide, nitrogen oxide, suspended particulates, and lead as well as emission limit values for SO_2, and solid particles, including smoke, soot, and dust, which may be different for new and existing plants. It also set requirements for progressively reducing the level of total emissions. A Royal Decree further modified the Decree 833/1975 in 1985 and introduced new limit and guide limit values for air quality standards.[52]

While following the interventionist approach of German and European air pollution control, Spanish regulations ignored the precautionary and technology-based principles. First, public authorities had considerable flexibility in applying emission standards. They could grant temporary exemptions from emission limit values to combustion plants, which used low quality fuels to overcome shortfalls in their energy supply or where 'local interests of social nature' (such as employment considerations) trumped environmental considerations – as long as air quality standards were not exceeded (Art. 52 Decree 833/1975). Second, Emission limit values were significantly lower and did not require the use of best available technology to reduce emissions. For example, the SO_2 emission limit values for combustion plants larger than 100 MW were nowhere near the limit of the LCP Directive for new plants. The SO_2 limit for combustion plants burning lignite was more than four times below the maximum limit value laid down in the Directive for plants smaller than 100 MW.[53]

There was a clear misfit between European and Spanish substantial air pollution standards. Moreover, as Spain had not implement the Industrial Plant Directive, Spanish air pollution control legislation was still wanting the precautionary and technology-based problem-solving approach also entailed in the LCP Directive, particularly regarding the application of BAT in reducing and monitoring emissions.

In order to bring Spanish emission limit values for combustion plants up to European standards, and to incorporate the total emission reduction levels for

---

[52]　*Royal Decreto 1613/1985 de 1 de agosto de 1985.*

[53]　But note that the LCP Directive allows for exceeding the SO_2 limit value if combustion plants burn indigenous fuel (Art. 5.2).

existing plants, Spain transposed the LCP Directive into Spanish law.[54] The Spanish text is almost a literal translation of the Directive. It was to simply replace the Decree of 1975 in all the parts, in which the latter did not conform to the Directive (*Disposiciones Adicionales, Primera*). The transposition of the LCP Directive did not only adapt emission limit values and introduce national emission ceilings to be reduced over a certain period of time. It had repercussions for monitoring air pollution, too. Spanish legislation had already prescribed a national monitoring network and detailed regulations on methods for measuring emissions and air quality standards. But Spanish measurement technology did not correspond to the 'best industrial measurement technology' required by Art. 13.2. of the Directive and taken over by Art. 14 of the Royal Decree (*mejor tecnología industrial de medición*, Bennett 1991: 62).

Unlike its Mother Directive, the LCP Directive was fully incorporated into Spanish legislation on air pollution. The 'automatic' suspension of conflicting provisions commanded by the Royal Decree of 1991 brought Spain in formal compliance with the LCP Directive.

But the considerable misfit between Spanish air pollution control regulation and the LCP Directive imposed considerable costs in practical application and enforcement, particularly after the five-year period expired, for which Spain had negotiated an exemption from SO_2 emission reductions.[55] First, between 1993 and 1998, Spain had to reduce its total SO_2 and NO_x emissions by 24 per cent. A further SO_2 reduction of 37 per cent and NO_x reduction of 20 per cent is envisaged between 1998 and 2003. In other words, Spain has been asked to cut its total SO_2 and NO_x emission by half till 2003. Second, Spain had to up-grade its measurement technology in order to meet the BAT requirements.

Spain has been among the highest SO_2 polluters in OECD Europe, together with the UK, Italy, and Germany. This is the combined effect of a reorientation of Spain's energy policy towards the consumption of domestic coal after the oil crisis in the early 1970s and the low quality of Spanish coal. Compliance problems with SO_2 emission standards of the LCP Directive for existing plants have been concentrated in Spain's two largest central power stations, As Pontes (Galicia) and Andorra (Aragón). The two plants account for more than half of the total SO_2 emissions in Spain. Andorra is considered the second most contaminating central

---

[54] *Real Decreto 646/1991, de 22 de abril de 1991*, BOE N° 99, 2.4.1991, modified by the *Real Decreto 18000/95 de 3 de noviembre 1995*, BOE N° 293, 8.12.1995, to comply with the Directive 94/66 extending the SO_2 emission limit values of the LCP Directive for new plants to plants burning solid fuel with a thermal output between 50 and 100 MW (they had initially been omitted). Besides, the Royal Decree set SO_2 emission limit values for oil refineries to be complied with from 2003 onwards.

[55] Art. 5 of the Directive granted Spain a temporary derogation from the full application of the SO_2 limits. In order to provide the generating capacity deemed necessary for its economic growth, Spain was allowed to authorize new solid-fuel burning plants bigger than 50 MW, which had to comply with less demanding standards. This derogation expired in 1999 and was only applicable to new plants that start operating before 2005.

power station in Europe. Following the requirements of the LCP Directive, the Spanish Ministry of Industry and Energy developed a National Plan, whose foreseen reductions went well beyond those required by the LCP Directive and could only be achieved by systematically applying BAT, or, by importing better quality coal. Both central power stations burn to 80 per cent indigenous high sulphur coal, and coal mining is the major source of socio-economic development in the areas where the central stations are located. Therefore, the Spanish administration was reluctant to take any measures to practically apply and enforce the ambitious emission reductions of the National Plan.[56]

Despite high levels of air pollution, domestic mobilization in favour of more effective air pollution control was low. Power plants are widely dispersed across the country and they are usually not located close to major urban centres. The alkaline nature of Spanish soils has mitigated the destructive effects of acid depositions. Problem pressures have arisen from soil erosions rather than corrosion (Font and Morata 1998: 209-210).

Yet, at the beginning of the 1990s, massive *Waldsterben* (forest dying) was observed in the neighbouring provinces of the Andorra plant. Local municipalities, environmental groups (among others Greenpeace), trade unions and political parties started to mobilize against the pollution caused by the central power station, which they considered as the source of forest dying. After a fierce public campaign, which included an environmental liability law suit (cf. Font 1996: 250-271), the management of the Andorra power station agreed to implement an environmental plan, in which it voluntarily committed itself to SO_2 emission reductions that superseded those required by the LCP Directive by 40 per cent. This 'over-compliance' with the Directive was motivated by the continued domestic mobilization of local citizen groups and environmentalists, which successfully linked the massive environmental deterioration to the pollution caused by the power station.

To what extent the Andorra plant has complied with its voluntary commitment, is difficult to tell in the absence of reliable measurement data (see below). According to an OECD report, Spain succeeded in significantly reducing its SO_2 emission over the last years. The reduction has been mainly achieved by substituting indigenous low quality coal through higher quality imports, not by using best available abatement technologies (OECD 1997). Spanish administration and industry alike have justified the non-application of BAT on the grounds of too extensive costs. Many of the existing combustion plants date back to the 1960s and 1970s and it would be economically prohibitive to upgrade their abatement technologies. In the absence of public subsidies, some companies threatened to close down their plants if they were forced to invest in new technologies, which increased the political costs for public authorities to enforce BAT requirements.[57]

---

[56]   Interview with the air pollution control unit of the Spanish Ministry of Environment, Madrid, 03/97.

[57]   Interviews with the environmental department of FECSA, the company, which runs most combustion plants in Catalonia, Barcelona, 03/97, and with the air pollution control unit of the Catalan Ministry of Environment, Barcelona, 03/97.

Air pollution control and enforcement has always been lax. Spanish administration has shown little interest in enforcing regulations on industries not complying with air pollution control regulations (Aguilar Fernandez 1994). Regulatory authorities have made considerable concessions regarding existing facilities, motivated by economic and employment concerns (OECD 1997: 80-82). Due to the uneven geographical distribution of measurement stations and a measurement technology, which is often not up to the standards, the Spanish monitoring system is not able to provide reliable data on air pollution for all areas. Cost considerations have prevented the necessary upgrading of the Spanish system for monitoring air quality in order to make it comply with the BAT requirement of the Directive. Monitoring air pollution falls under the responsibility of the Spanish regions, of which many lack the financial resources to invest in better monitoring technology, additional manpower and expertise. Catalonia, for instance, invested more than EUR 421,000 in the creation of an integrated air pollution monitoring network (*Red de Vigilancia y Prevención de la Contaminación Atmosférica*), which still does not cover the whole region and the number of polluters it can measure is limited (Choy i Tarrés 1990: 204). While air pollution monitoring has improved over the last years, the Spanish system – depending on the region – is not always up to 'the best industrial measurement technology' required by Art. 13.2. of the Directive and Art. 14 of the Royal Decree of 1991 (*mejor techología industrial de medición*; cf. OECD 1997; Instituto para la Política Ambiental Europea 1997).

In sum, while compliance with European air pollution control policies has improved, the BATNEEC requirements are not systematically enforced either on existing combustion plants or on monitoring networks. An operationalization of the concept is still missing but it might finally emerge within the implementation of the Integrated Pollution Prevention and Control Directive passed in 1996 (see above). The IPPC Directive, which builds on the Industrial Plant Directive, stipulates that permits for industrial plants must set emission limit values based on what is actually achievable through the use of best available technology but under economically and technically viable conditions. Following the requirements of the Directive (Art. 16.2), the European Commission organizes horizontal exchanges of information on BAT in Technical Working Groups, which consist of experts from the member states, industry and environmental organizations. The result of these exchanges are laid down in so called BAT Reference documents (BREFs), which the member states have to incorporate in their environmental permitting procedures. BREFs are produced for each industrial sector listed in Annex 1 of the Directive. The European IPPC Bureau in Seville coordinates the work of the Technical Working Groups. Spain participates in about two-thirds of the working groups. Emulating the European example, the Spanish Ministry of Environment established seven working groups in 1998, bringing together representatives of the central state administration, the Autonomous Communities, industry and scientific experts to define BAT standards for the various sectors of Spanish industry. Whether those standards will be ultimately enforced, remains to be seen.

Germany: Complete Fit

The two Directives on air pollution control were modelled on the German technology oriented, precautionary, emission based problem-solving approach in combating environmental pollution. The Federal Immission Control Act (*Bundesimmissionsschutzgesetz/BimSchG*) of 1974 served as the blueprint for the Industrial Plant Directive while the Large Combustion Plant followed the German Large Combustion Plant Ordinance (*Großfeuerungsanlagen-Verordnung/GFAV*). As a result, the two Air Pollution Directives did not produce any significant adaptational pressure for Germany.

The German regulatory structures for combating air pollution by industrial plants date back into the late 19th century. From the very beginning, they entailed a legalistic and interventionist approach, where regulatory requirements are legally imposed and leave little flexibility and discretion for the administration to adopt control standards that correspond to particular circumstances, such as the state of the local environment or the economic situation of industry (Héritier, Knill, and Mingers 1996: 39-55). The General Trade Regulations (*Allgemeine Gewerbeordnung*) entitled Specific Trade Offices (*Gewerbeämter*) to restrict the operation of industrial plants in the public interest. The Federal Act on Combating Air Pollution (*Luftreinhaltegesetz, LRG*) of 1959 extended the provisions of the General Trade Regulations. It set a variety of air quality standards. Moreover, it increased the type of industrial plants requiring a permit and which had to apply the best available technology in their production (Boehmer-Christiansen and Skea 1991). In 1974, finally, the Federal Immission Control Act replaced the General Trade Regulations.[58] The new law defined precaution, the-polluter-pays, cooperation, and emission abatement on the basis of the best available technology as the basic principles of air pollution control. In accordance to these principles, the Federal Immission Control Act has regulated, among other things, the procedures for the authorization of the establishment as well as the modification of industrial plants that are likely to have a negative impact on the environment (§§ 10, 15. 16). Authorization is to be granted if 1) the plant will not cause any negative impact on, or dangers for, the environment; 2) all appropriate preventive measures to minimize air pollution were taken, including the application of the best available technology in reducing emissions, 3) all legal obligations are complied with (emission limit values, air quality and production standards §§ 5, 6).[59] A Federal Ordinance determines the plants that

---

[58] *Gesetz zum Schutz von schädlichen Umwelteinwirkungen durch Luftverunreinigungen, Geräusche, Erschütterungen und ähnliche Vorgänge (Bundes-Immissionsschutzgesetz, BImSchG), in der Fassung vom 14. Mai 1990,* BGBl I: 880.

[59] Industrial plants not subject to prior authorization have to comply with the same requirement (§ 22).

are subject to authorization.[60] The Federal Immission Control Act has also entailed detailed provisions on the monitoring of air pollution and the implementation of air pollution plans to reduce emissions in areas which supersede the legal air quality standards (§47). Finally, the Federal government has been entitled to adopt ordinances (*Rechtsverordnungen*) and administrative directives (*Verwaltungsrichtlinien*) to prescribe product standards, air quality standards and emission limit values (§ 7).

After the Federal Immission Control Act had been enacted, an administrative directive (*TA Luft*) regulated the details of its application and enforcement.[61] The *TA Luft* provided a definition of 'air pollutants', specified authorization procedures and set emission limit values for various types of industrial installations. But as an administrative directive, it did not have any external effect, that is, it only obliged the administration. While administrative directives allow for some flexibility in the enforcement of strict uniform emission standards, they do not set binding law (Kloepfer 1984: 261).[62] Third parties, such as citizens or courts, are not subject to administrative directives. Since industry was concerned that the non-binding character of the *TA Luft* could result in the non-uniform application of emission limit values across Germany, the federal government, with the consent of the *Bundesrat*, enacted a federal ordinance in 1983.

---

[60] *Vierte Verordnung zur Durchführung des Bundes-Immissionsschutzgesetzes (Verordnung über genehmigungsbedürftige Anlagen, 4. BImSchV), vom 24. Juli 1985, BGBl. I: 1586.*

[61] *Technische Anleitung zur Reinhaltung von Luft (TA Luft) vom 8. September 1964, Gemeinsames Ministerialblatt vom 14. September 1964: 433.*

[62] The jurisdiction of the German courts, however, is ambivalent on this issue. Some courts fully denied administrative directives any legally binding effect (*Voerde-Urteil des Bundesverwaltungsgerichts zur TA-Luft 1978, BVerwGE 50, 250*). Others acknowledged a limited legal effect if administrative directives specify existing norms; but they could not be subject to judicial scrutiny (*Buschhaus-Urteil des Oberverwaltungsgerichts/OVG Lüneburg zur TA-Luft 1985, Deutsche Verwaltungsblätter 1985: 1322 ff*; cf. *OVG NW, Deutsche Verwaltungsblätter 1988: 152-153*). In a more recent ruling, the Federal Administrative Court interpreted administrative directives as 'norm concretizing' (*normkonkretisierend*) which implicates a legally binding effect for administrative courts (*Wyhl-Entscheid des Bundesverwaltungsgerichts zum Atomrecht, BVerwGE 72, 300*). German administration usually applies administrative directives as if they were law (Reinhardt 1992). The European Commission, however, has insisted that legally non-binding administrative directives are not sufficient to effectively implement European Law, most of all because they have not to be made public (Commission of the European Communities 1991: 276). The Commission initiated several infringement proceedings against Germany, because the latter had implemented several water and air pollution control Directives by means of administrative directives (cf. Börzel 2002b: 158-159).

The Large Combustion Plant Ordinance (*Großfeuerungsanlagen-Verord-nung/GFAV*)[63] imposes nationally uniform emission standards based on Best-Available Technology (BAT) and precautionary action. It reflects the technical understanding of environmental protection, which has prevailed in Germany. The risks and damaging impacts of air pollution are to be countered by plant-related measures. Even plants in less stressed or unpolluted areas have to comply with the emission limit values. According to the polluter-pays-principle, emissions are to be abated directly at source by means of highly developed environmental protection technology. The *GFAV* imposed a 60 per cent across-the-board reduction in sulphur dioxide for all combustion plants (new and existing) with a rated thermal output of 50 MW and more, to be achieved within five years. Strict emission limits were also set for other major polluters, including nitrogen oxide, carbon oxide, halogen compounds, dust, and carcinogens. Although the controls were specified in terms of emission limit values, the standards were so stringent that plants could only meet them by using expensive flue-gas desulphurization technology. Since the standards also applied to old plants, the *GFAV* required substantial retrofitting of the new abatement technology to existing power stations. Finally, the *GFAV* prescribed detailed monitoring requirements.

The Large Combustion Ordinance met fierce resistance form German industry, which fought the proposal before administrative courts on the grounds of the principle of proportionality, arguing that compliance costs were disproportionate in relation to the expected benefits. German business feared that higher electricity prices would lead to competitive disadvantages. Its judicial appeals failed, however, and public pressure was too high on the government to yield to economic opposition. But German policy-makers managed to make the strict regulations more palatable by stressing technologically oriented solutions that would enhance German industrial competitiveness. Germany also sought to harmonize its stringent standards at the European level, a strategy which German industry finally supported as the second-best option (Weale et al. 2000: 387-388).

Once the German Large Combustion Plant Ordinance had been passed, it was effectively applied and enforced (cf. Lenschow 1997). Between 1983 and 1988, the practical application of the *GFAV* had led to a reduction of sulphur dioxide emissions by 88 per cent and of nitrogen oxides by 76 per cent (Weale et al. 2000: 304). Since then, Germany has achieved further reductions. The intensity of SO_2 and NO_x emissions is 65 per cent and 50 per cent below OECD averages (OECD 2001: 38-48). The implementation success is largely attributed to the corporatist framework of economic policy-making which helped to coordinate manufacturers to produce pollution control equipment as part of the counter-cyclical economic policy (Boehmer-Christiansen and Skea 1991). Public investments in environmental research and development rose from a share in total investments of 3.5 per cent in 1980 to 4.4 per cent in 1984 to 5 per cent in 1991 (Weidner 1995: 74),

---

[63] *Dreizehnte Verordnung zur Durchführung des Bundes-Immissionsschutzgesetzes (Verordnung über Großfeuerungsanlagen), vom 22. Juni 1982*, BGBl. I: 719.

which gave rise to a growing pollution control industry, some have referred to as the 'eco-industrial complex' (Weale et al. 2000: 270).

Since the two European Air Pollution Control Directives were modelled after German legislation, Germany has not faced any compliance problems. Particularly with regard to the LCP Directive, German requirements were both more extensive and more demanding. First, the *GFAV* covered seven and not only three substances as the Directive did. Second, the *GFAV* prescribed uniform emission limits for all plants, whereas the Directive linked the requirements for old plants to the level of national emission ceilings – as long as the average of total emissions did not exceed a certain limit, individual plants did not need to take any reduction measures. Third, the *GFAV* requirements for the application of the best available technology were more demanding. Due to the uniform and strict emission limit values, all industrial plants had to fit flue-gas desulphurization and selective catalytic reduction technology. The Directive required flue-gas desulphurization only for installations with a rated thermal output of 300 MW while smaller plants had to use low sulphur fuel. Low NO_x burners were sufficient to meet the European NO_x limits (Boehmer-Christiansen and Skea 1991: 236-238). In short, Germany had no difficulties in meeting the reduction targets set by the LCP Directive. In fact, it was several years ahead of the European timetable.

To conclude, while the two European Directives on combating air pollution fit the corresponding German policies, they have exerted considerable pressure for adaptation on Spanish air pollution policies. The required modernization of Spanish industrial plants imposes high costs which neither industry nor public authorities have been inclined to bear. As a result, the Industrial Plant Directive has not been implemented at all. The LCP Directive has simply been absorbed into the Spanish system leaving substantial parts non-applied. The Spanish administration has shown little willingness to deal with the considerable adaptational costs incurred by cutting down the Spanish total SO_2 and NO_x emissions and by bringing the Spanish monitoring system up to BAT standards. The Commission, in turn, has never reprimanded Spain for not complying with the two Directives. Only when societal actors and municipalities formed a powerful coalition pressuring both central state administration and industry to comply with existing air pollution standards, Spanish industry and public authorities started to become active. Since then, Spain has considerably reduced its SO_2 emissions. But it was industry that ultimately responded to domestic pressure for adaptation by committing itself voluntarily to emission reductions well beyond the requirements of the Large Combustion Plant Directive.

Environmental Impact Assessment: Misfit in Spain and Germany

The Policy and its Development

The 5th Environmental Action Programme (1992-1997) indicated a change in European environmental policy-making. The principle of shared responsibility between state authorities, industry, consumers, and the general public shall complement the traditional command-and-control approach. The Directive on Environmental Impact Assessment[64] is one of the first policies to implement this new approach. Environmental impact assessment (EIA) follows a precautionary problem-solving approach. It is based on the idea that the prevention of environmental damage requires detailed information on the impact of particular projects and activities. The EIA Directive constitutes an instrument of procedural regulation, which assesses in a systematic and cross-sectoral way the potential impact of certain public and private projects on the environment. The policy also contains a strong element of public participation. The basic principle of the EIA Directive is that any project which is likely to have significant effects on the environment by virtue *inter alia* of its nature, size or location is subject to an environmental impact assessment prior to authorization by the competent authority (cf. Haigh 2001: 11.2-1-11.2.-4). Projects listed under the 21 headings in Annex I of the Directive must be made subject to an EIA. Projects listed under the 13 headings in Annex II, by contrast, shall be made subject to an EIA only if they are likely to have a significant effect on the environment. The member states have to decide on a case by case basis, and/or by reference to threshold criteria whether a project is made subject to an assessment or not. Direct and indirect impacts on the environment have to be assessed with regard to several categories, including human beings, fauna, flora, soil, water, air, climate and landscape, the interaction between the first two groups, material assets and cultural heritage. The developer of a project has to provide information on the characteristics of the project, its environmental impacts, the measures envisaged in order to avoid, reduce and remedy significant adverse effects, and the data required to identify and assess the main potential effects of the project. Annex III sets out guidance on the type of information to be provided. The member states can establish a procedure, in which the competent authority and the developer informally agree upon the precise scope of the required information ('scoping procedure'). This information shall be forwarded to public authorities likely to be concerned by the project. Authorities must also make available to the public the information, which they gathered to assess the environmental impact of a project. The public concerned must have the possibility to make allegations. Member states have to specify the requirements for the provision of information and for public hearings. Finally, the competent authorities have to consider the information, together with the public allegations made, and to decide on the positive or negative environmental impact of a solicited project. Authorities have to

---

[64]   85/337/EEC; OJ L 175, 5.7.85.

inform the public about their final decision as well as on the reasons and considerations on which they based their decision.

The Directive was amended in 1997.[65] The revisions were to address some major problems that the Commission had identified in its evaluation report on the implementation of the EIA Directive (Commission of the European Communities 1993a). The list of Annex I projects was expanded and the treatment of Annex II projects clarified; the member states have to determine the requirements for an EIA of Annex II projects either through a case-by-case examination or through threshold criteria as outlined in Annex III of the amended Directive. The assessment of the interaction between all the factors listed in the Directive was explicitly required. Moreover, the minimum requirements for an EIA study were extended, including an outline of the main alternatives studied by the developer and an indication of the main reasons for her choice. The 'scoping' procedure, which some member states had already implemented, was formally incorporated in the Directive. Finally, the Directive established a single procedure for projects covered by both the IPPC and the EIA Directives.

Unlike the Drinking Water and the Air Pollution Control Directives, the EIA Directive did not originate in a regulatory contest among the member states but was an initiative by the Commission, which in turn had been inspired by a US-regulation in the National Environmental Policy Act of 1969 (cf. Staeck, Malek, and Heinelt 2001: 33-35). The Commission formally announced its intentions of proposing an 'environmental impact assessment' as practiced in the US in the 2nd Environmental Action Programme (1977-1981). The formal proposal, which the Commission presented in 1980, received little support by the member states. Most of them considered environmental impact assessment as a useful concept. But they strongly disagreed on the type of project to be assessed, the assessment criteria, and the scope of public participation. National governments were concerned about the legal and institutional adjustment costs, which the incorporation of the procedural regulations would incur. Even the UK, whose regulatory tradition was largely in line with the procedural approach of the Directive and which had practiced EIA on an albeit voluntary basis since the 1970s, was sceptical. The British government feared that the EIA procedure could empower the courts vis-à-vis public authorities (Haigh 2001: 11.2-6). Denmark had similar concerns with respect to projects authorized by an Act of Parliament. The conflictual nature, which the EIA could inflict on the authorization process with its obligatory and more demanding regulations was also contrary to the British regulatory style, which favoured consensus and informal agreement (Héritier, Knill, and Mingers 1996: 295). British objections were overcome by a series of concessions negotiated in over 40 meetings of the Council working group on environmental affairs. The amendments to the initial Commission proposal reduced the list of projects subject to a mandatory EIA in Annex I. Member states could also make exemptions without explicit approval of the Commission. Moreover, public authorities would have considerable discretion

---

[65]   97/11 EC; OJ L 73/5, 14.3.97.

with regard to the type and scope of information to be provided for the EIA. To accommodate the Danish concerns, projects that were adopted by national legislation were exempted from EIA requirements. The Directive was finally adopted in June 1985 and the member states had two years to incorporate it into their domestic legislation. While the amendments of the original Commission proposal made European and British regulations by and large fit, it strongly contradicted regulatory structures in Germany, whose media-specific environmental legislation lacked any comparable arrangements (see below). The German government considered the Directive as superfluous since similar assessment methods were already applied in sector-specific planning procedures. But while Germany was aware that the implementation of the new instrument would require substantial modifications of its environmental legislation and administrative practice, it did not voice much opposition in the decision-making process. 'It was simply that no-one in Germany had expected the Directive ever to be adopted after the never-ended negotiations' (Héritier, Knill, and Mingers 1996: 296).

Spain: Diffuse Pull and Reluctant Push

Before the EIA Directive, there was no comprehensive environmental impact procedure in Spain. But like in other member states, different sectoral environmental regulations required the assessment of certain environmental impacts of a planned project. The most important predecessor of EIA was the Regulation on Irritating, Unhealthy, Harmful, and Dangerous Activities.[66] This administrative decree of 1961, partly modified in 1965, required a sort of EIA for so called 'classified activities'. Before local authorities granted the license required for a new classified activity, or the modification of an already existing one, the impact of such an activity on environmental conditions relevant to public health had to be assessed. Similar to the EIA Directive, the promoter had to provide information on the characteristics of the activity, its potential repercussions on the environment, and the corrective measures to remedy such adverse effects. The procedure also included a period of public information of ten days, in which the public concerned could make allegations. The regional environmental authority assessed the environmental impact of a classified activity. In case of a negative assessment, the local license had to be denied.

The notion *'evaluación de impacto ambiental'* (EIA) appeared for the first time in Spanish legislation in a circular of the Ministry of Industry on the law on air pollution control in 1976.[67] The circular explicitly required an EIA for the installation of new industrial plants and the modification of existing industrial plants (Art. 2, 1a) which, however, only applied to the aspect of air quality. In a similar vein,

---

[66] *Reglamento de Actividades Molestas, Insalubres, Nocivas y Peligrosas (RAMINP) de 30 de noviembre 1961 (Decreto 2414/1961;* BOE N° 292, 7.12.1961).

[67] *Orden de 18 de octubre 1976, Prevención y corrección de la contaminación atmosférica de origen industrial,* BOE N° 290, 3.12.1976.

some sectoral legislation on water, coastal lines, waste, soil, and mines prescribed or recommended the realization of studies assessing the impact of these activities on the environment.[68]

While environmental impact assessment was not a new concept, Spanish regulations did not entirely fit the requirements of the EIA Directive. They lacked the explicit precautionary approach entailed in the Directive. Moreover, the provisions for public participation and transparency in administrative procedures conflicted with the Spanish administrative practice of closure and secrecy. The Spanish requirements for information to be provided by the developer were much less demanding. The period of public information and consultation was shorter, too. Requirements for corrective measures were lax to non-existent. Cross-media effects were not systematically considered either. Finally, a project was only assessed with respect to its negative effects on public health, not on the environment as such.

Due to the misfit between European and Spanish EIA regulations, Spain opted for a proper law to implement the EIA Directive, rather than integrating it into the different pieces of sectoral legislation. Whereas the EIA Directive had already been transposed in 1986,[69] the national transposition law was only executed two years later, in 1988.[70] The Spanish EIA procedure by and large conforms to the procedural requirements of the Directive.

1) The promoter sends a summary of the project to the environmental authority. 2) The environmental authority *may*[71] consult other administrative bodies and actors presumably affected by the project, who have 10 days to state their opinion. 3) The opinions have to be transmitted to the promoter within 20 days, together with an indication of the contents required for the Environmental Impact Study (EIS). 4) The promoter prepares the EIS and sends it, together with the other documents required for authorization, to the authorizing body (competent authority), which forwards the EIS to the environmental authority in charge of the impact assessment. 5) A note is published in the Official Journal of the central state or the region. If regular authorization procedures do not provide for public information, the 30 days period stipulated in the Directive is initiated. 6) Any allegations made during the public information are sent to the promoter, who shall respond to them within 20 days. 7) The environmental authority has 30 days to assess the EIS, to issue the EIA declaration, and to send it to the competent authority. If there is a dissent between the competent and the environmental authority, the government decides. 8) The EIA declaration is published in the Official Journal. Without a positive EIA, authorization must not be

---

[68] *Real Decreto 2994/1982, Restauración de espacios naturales afectados por actividades extractivas*, BOE N° 274, 15.1.1982; *Ley 29/1985 de Aguas*, BOE N° 189, 8.8.1985 and BOE N° 243, 10.10.1985; and *Ley 22/88de Costas*, BOE N° 181, 29.7.1988.

[69] *Real Decreto Legislativo 1302/1986, de 28 de julio, de evaluación de impacto ambiental*; BOE N° 155, 30.6.1986.

[70] *Real Decreto 1131/1988, de 30 de septiembre*, BOE N° 239, 5.10.1988.

[71] Art. 6.1 of the Directive stipulates that the member states 'shall' take the measures necessary to ensure that these authorities are consulted.

granted. If a project is executed without the necessary EIA, the environmental authority can demand the suspension of the project. The same applies in cases in which the promoter is guilty of fraud, misinformation, and non-implementation of corrective measures. Monitoring compliance lies in the responsibility of the competent authority.

But transposition was neither complete nor entirely correct. The Spanish EIA regulations only included six of the Annex II projects, including large dams, initial afforestation, airports for private use, marinas, and extraction of coal, and lignite as well as minerals by open-cast mining (Anexo 1, RD-L 1302/1986; Anexo 2, RD 1131/1988). While those six were made subject to a mandatory EIA, all the other projects listed in Annex II were completely omitted. The Spanish government claimed that Art. 4 of the Directive left it to the discretionary judgement of the member states whether the undertaking of EIA for Annex II projects should be mandatory (Commission of the European Communities 1993a: 231). The Commission contended that member states had to make decisions on whether Annex II projects were subject to an EIA either on a case-by-case basis or by defining clear threshold criteria. The disagreement resulted in the opening of an infringement proceeding in 1990. After a Reasoned Opinion had been sent in 1992, the Spanish government finally agreed to remedy the matter till 1994. In 1995, Spain still had not amended its legislation. The Spanish Ministry of Environment declared its intention to correct the legislation with the transposition of the revised EIA Directive, which was sufficient for the Commission to suspend the proceeding once again. The Spanish government prepared more than 15 proposals of which none materialized due to the resistance of various national ministries whose competencies were affected by the omitted Annex II projects. After the deadline for transposing the revised EIA Directive had expired, the Commission finally referred the case to the European Court of Justice in December 1999,[72] where it is still pending. It also opened new proceedings for the non-transposition of the revision of the EIA Directive. The case reached the ECJ in September 2000,[73] but the Commission withdrew when the Spanish government prepared a draft law to remedy the situation. The issue was settled in October 2000, when Spain validated the draft law.[74]

Some of the Autonomous Communities enacted their own EIA regulations going beyond the central state transposing legislation, by including several of the Annex II projects (Madrid, Extremadura, Valencia, Aragón, Cantabria, Baleares), adding additional projects not subject to the Directive (Catalonia and Galicia), or introducing a more simple environmental assessment for certain projects neither

---

[72] C-1999/474, 14.12.1999.

[73] C-2000/342, 15.9.2000.

[74] *Resolución de 19 de octubre de 2000, del Congreso de los Diputados, por la que se ordena la publicación del Acuerdo de convalidación del Real Decreto-ley 9/2000, de 6 de octubre, por el que se modifica el Real Decreto Legislativo 1302/1986, de 28 de junio, de evaluación de impacto ambiental*, BOE N° 256. 25/20.2000.

subject to Annex I nor II (Cantabria, Canaries, Navarre, Extremadura, Madrid, Castilla-León). As a result, the practical application of the EIA Directive varies considerably from one Autonomous Community to another (López Taracena 1995).

The (incomplete) transposition of the EIA Directive gave also rise to some constitutional conflicts over competencies between the central state and the Autonomous Communities. The Basque Country, for instance, claimed that several regulations of the Royal Decree 1131/1988 would interfere with its sphere of competencies.[75] The designation of the environmental administration as the competent authority for the EIA (Art. 4) and the monitoring of the implementation of corrective measures (Art. 25) as well as the designation of the competent authority for the resolution of conflicts between the administrative unit issuing the EIA and the one which grants the overall authorization of a project (Art. 20) would clearly fall under the exclusive responsibility of the Autonomous Communities. The Basque Country also denied the central state the competence to exempt a specific project in whole or in part from the requirement of an EIA (Art. 2b, 3), if this project was to be realized on Basque territory. The central state administration defended these regulations in light of its competence to set basic legislation in order to provide a uniform framework for the application of the EIA Directive in Spain. It also emphasized that the central state government or the national Parliament would only decide on exemptions of projects that were of general interest, i.e. fell under the exclusive competence of the central state. The Constitutional Court dismissed the case in January 1998 confirming the central state competence to decide on the EIA for projects, which are authorized by the central state. As a result, the Autonomous Communities are deprived of a potential veto right on projects of general interest realized within their own territories.

While transposition was already incomplete, the practical application of the EIA Directive has not been effective either. Administrative changes have been limited. The EIA procedure was incorporated into the existing administrative procedures for the implementation of *RAMINP*. These procedures were merely formally adjusted to the requirements of the European EIA procedure. Consequently, public authorities have not dedicated sufficient manpower and expertise to ensure the good quality of EIS, its appropriate assessment, and the enforcement of corrective measures, especially if the relatively short time period for the assessment and the cross-media requirements of the Directive are considered. EIAs are often carried out 'in the old way', i.e. in the way former national EIA under *RAMINP* used to be conducted. As a result, the quality of the EIS tends to be low (Commission of the European Communities 1993a). Alternatives are not really discussed; the 'zero alternative' of not realizing the project is not considered at all (Escobar Gómez 1994). Public and industrial are hardly turned down (Escobar Gómez 1994). The local authorities in charge of monitoring often cover up for the

---

[75] *País Vasco, Conflicto n° 263/1989.*

'circumventing' of authorization procedures in order not to suffer disadvantages in the competition over jobs and economic investments.[76]

Project promoters have little incentive to comply with EIA requirements, either. The elaboration of adequate EIS requires expertise, in which many enterprises have not been willing or able to invest. An EIA can increase both the costs and the time scale of a project, the former by an estimated 5-10 per cent, and the latter by a couple of months delay (Commission of the European Communities 1993a: 241). Since public authorities often appear reluctant to apply and enforce the EIA obligations, many projects proceed without any authorization at all, or are only asked for when they were already carried out.

In general, the effectiveness of EIA implementation is rather low in Spain. In a number of cases, however, EIA has turned out to be more than mere 'rubber stamping' of licenses and permits. Societal actors have repeatedly mobilized against the ineffective application of the EIA Directive, trying to pull the policy down to the domestic level. If correctly applied, the new EIA procedure strengthens their position in the authorization process of public and private projects.[77] First, the period of public information is longer (30 instead of ten days) giving societal actors more time to prepare their allegations. Second, the information on the environmental impact of a project is more detailed and has to take into account the cross-media effects – not only on public health but also on the environment as a whole. This gives the public concerned a broader basis to make allegations. Finally, the European Commission is a powerful ally to whom societal actors may turn with their complaints about ineffective implementation instead of making appeals to the Spanish courts, which tend to be time consuming, costly, and often unsuccessful.

Environmentalists and citizen groups, often in coalition with local authorities, have increasingly made use of the participatory opportunities provided by the EIA to voice their opposition against the authorization of public and private projects. In a number of cases domestic mobilization stopped a project or, far more likely, led to corrective measures imposed on the promoter. A case study conducted at the Universitat Autònoma de Barcelona on the authorization of nine different projects subject to EIA – eight public, one private – shows that domestic mobilization can make a difference. In all cases but one, environmental groups, often together with the local governments concerned, strongly opposed the project. In three cases, they made a formal complaint to the Commission. Three of the public projects were not authorized due to a negative EIA, and the authorization of another three projects was made conditional on substantial corrective measures. Hence, in two thirds of the cases, the EIA had a significant impact on the outcome of the project (Font 1996).

---

[76]   Interview with the EIA unit of the Catalan Ministry of Environment, Barcelona, 03/97.

[77]   Interviews with ADENA-WWF and AEDENAT, Madrid, 03/97, and with DEPANA, Barcelona, 03/97.

Statistical data confirm that the EIA Directive has stimulated domestic mobilization. Denouncements and petitions submitted to the Spanish Parliament with respect to infringements of the EIA regulations account for about 40 per cent of the total number in the environmental sector.[78] Together with the Habitat and Wild Bird Directive (nature), EIA also represents the highest number of Spanish complaints to the Commission (about 30 per cent). Among the member states, Spain faces one of the highest numbers of complaints filed against the ineffective implementation of the EIA Directive (Commission of the European Communities 1996). The complaints resulted in the opening of four infringement proceedings for not properly applying the EIA Directive (1996, 1997, 1998, and 1999). Of the four cases, one was referred to the ECJ[79] and two are still pending at the stage of a Reasoned Opinion.

Despite some success, the implementation of the EIA Directive in Spain points to an important caveat with respect to the effect of this new policy instrument. Empirical evidence shows that *if* societal actors mobilize against the authorization of a project, EIA strengthens their position vis-à-vis public authorities by enabling them to prevent a project altogether or, at least, to impose corrective measures. Yet, societal actors, particularly at the local level, have only limited resources to mobilize in the first place. They often lack the necessary manpower, expertise, and money to oppose the non-application of EIA regulations such as a cross-media assessment of the potential environmental impact or the elaboration of corrective measures. In other words, societal actors often do not have the necessary resources to exploit the opportunities for mobilization offered by EIA. As a result, societal mobilization to pull EIA down to the domestic level is concentrated on so called NIMBY issues ('not in my backyard'), which seriously affect the 'backyard' of a larger group of people who then decide to join forces in mobilizing against public authorities.

Germany: Combined Pull and Push

The EIA concept forms part of media-specific authorization procedures in Germany. The environmental harmful effects of a project are assessed for each media individually and by different authorities. The EIA is to ensure that the project will comply with the legal standards that apply. Depending on the authorization procedure, affected parties can make allegations in public hearings before the consent decision is made (see below). The media-specific approach is deeply rooted in the German administrative culture with its belief in the technical superiority of bureaucratic organization through specialization as developed by Max Weber. The organization of the environmental administration both at the Federal and the *Länder* level follows a pattern of departments, sub-departments and highly specialized sections (*Referate*), which are responsible for the different media. These traditional

---

[78] *Medio Ambiente en España 1994, 1996, 2000* (Annual Reports of the Spanish Ministry of Environment).

[79] C-2001/227.

bureaucratic structures have prevented the emergence of an integrated approach to environmental problems.

Since Germany had no tradition of integrated environmental impact assessment, the EIA Directive has exerted considerable pressure for adaptation on its legal and administrative structures. The sectoralization of environmental regulation contradicts the cross-cutting demands of environmental management, in which problem solutions have to be seen in a holistic and integrated way. First, the effective implementation of the EIA Directive requires the coordination of various authorization procedures dealing with the different environmental media, such as water, air, soil, which fall into the responsibility of different authorities located at different levels of government. The designation of a lead agency that is in charge of the overall coordination and control of EIA procedures also implies a centralization of administrative competencies, since administrative processing of a project is not coordinated in either legal terms or practical performance (Knill 2001: 143-145).

Second, the procedural nature of the EIA Directive challenges the German problem-solving approach with its emphasis on rigid substantive regulations set by highly media-specific laws, which leave public authorities only limited discretion in application and enforcement. The EIA Directive, by contrast, not only requires the competent administration to cooperate with other authorities but to make an assessment decision about the environmental impact across the different media. Its procedural requirements do not provide any assessment criteria in form of regulatory standards to be applied by the authorities.

Third, while German provisions for public participation in authorization procedures are more extensive than in many other countries, the EIA Directive is still more demanding. The German legal order is strongly oriented towards the protection of the subjective rights of the individual. The administrative law reflects this prevailing concern by legally protecting the rights and interests of the *directly affected* individual (Nolte 1994). German legislation does not provide general rights to access project or planning applications, make representations and to object administrative decisions. Only those whose interests are directly affected are entitled to participate in administrative procedures and to litigate against their outcomes. The necessity of a pluralistic participation of the public is not acknowledged as a value in itself, for instance because it would allow to draw on multiple sources of information in assessing the environmental impact of a project. Thus, German provisions for public involvement rest on a defensive and reactive form of participation, where a limited group of potentially affected citizens are entitled to exercise their claims.[80] Broad public involvement in the EIA procedure not only conflicts with German administrative tradition, it is also considered as a

---

[80]   There are some exceptions though. The Nature Protection Law allows for the formal participation of environmental groups in the preparation of regulations as well as programmes and plans (§ 29 *Bundesnaturschutzgesetz vom 6. August 1993*, BGBl. I 880). In general, however, Germany is opposed to the concept of standing rights for public interest groups.

challenge to the international competitiveness of German industry, commercial freedom and confidentiality. Business associations argued that public participation would delay licensing processes and render projects more expensive due to the corrective measures imposed (Héritier, Knill, and Mingers 1996: 296; Héritier et al. 1994: 308-310).

Notwithstanding these significant incompatibilities, German administrators have considered the transposition of the Directive unnecessary claiming that existing legislation already covered all the provisions of the Directive and even went beyond some requirements. 'We do not see the point of changing our legislation only to put the formal label "Environmental Impact Assessment" on something we have been practicing long before the Commission discovered the EIA'.[81] While the Commission made clear that it expected Germany to formally transpose the Directive, the horizontal and vertical fragmentation of environmental competencies delayed the legislation process (Staeck and Heinelt 2001: 61-63). Agreement was difficult to achieve among the various ministries at the federal and regional level whose responsibilities were affected and most of which were strictly opposed to enacting a separate EIA law. As an official of the Federal Ministry of Environment explained: 'The implementation of the *UVG* required amendments of eleven different federal laws, which raised the opposition of eleven different administrations both at the federal and the *Länder* level. And we had to cope with concerns of German industry, too. Actually – we had almost everybody against us'.[82] Since some of the projects listed in the EIA Directive fell under the competencies of the *Länder*, they were required to adapt their regulatory structures, too. Only Bavaria, Baden-Württemberg and North-Rhine-Westphalia already possessed their own EIA laws, which they had to change. The other *Länder* had to establish EIA guidelines, amend technical legislation or adapt administrative procedures.

Since Germany delayed the transition of the EIA Directive for two years, several environmental organizations claimed direct effect demanding an EIA for the authorization of Annex I projects after the transposition deadline had expired in July 1988. They filed several court cases against projects authorized without an EIA, and launched a series of complaints to the Commission. Following the complaint of a German environmental organization (*Öko-Institut*), the Commission finally opened infringement proceedings in 1990 against the authorization of a power plant without EIA. The German government admitted that the EIA Directive had not been transposed yet, but contended that existing legislation already complied with European requirements and denied the direct effect of the EIA Directive. The Commission referred the case to the ECJ in 1992. The ECJ dismissed the case for failure of the Commission to specify where the requirements of the German authorization procedures did not comply with the EIA Directive.[83] Part of the German

---

[81] Interview with the EIA unit of the Bavarian Ministry of Environment, Munich, 09/98.

[82] Interview with the EIA unit of the Federal Ministry of Environment, Bonn, 10/97.

[83] C-1992/431, 11.8.1995, cf. Kunzlik 1996; Spindler 1994.

administration took the ECJ ruling as a clear indication that German regulations complied with the EIA Directive.[84] While the German environmental minister, Klaus Töpfer, accepted the direct applicability of the Directive, public authorities often refused to insist on an EIA (Héritier, Knill, and Mingers 1996: 298).

After four years of resistance, Germany finally yielded to the combined pressures from below and from above and formally transposed the EIA Directive in 1990. German policy-makers opted for the implementation through a separate law, the Environmental Impact Assessment Act (*Umweltverträglichkeitsprüfungsgesetz/UVPG*),[85] which did not establish a separate EIA procedure but integrated EIA into existing authorization procedures. The first part of the *UVPG* established the procedural framework for all authorization procedures. It stipulated the aims, areas of application, and regulations covered by the procedure, allowing for more detailed and far-reaching regulations in the media-specific legislation of *Bund* and *Länder*. The second part of the *UVPG* specified the necessary amendments required in eleven media-specific laws at the national level.[86] The German EIA Act by and large complies with the procedural requirements of the Directive.

> 1) The promoter informs the competent authority of the planned project. A 'scoping' process determines the information, which the promoter shall provide. The competent authority consults the promoter and interested parties (other public authorities, experts, environmental groups), decides the methods, content and scope of the EIA, and informs the promoter accordingly. The authority *may* also require the promoter to consider the environmental impact of potential alternatives to the project. 2) The competent authority collects the information and makes it available to affected parties, which can make observations in a public hearing. 3) The competent authority summarizes the expected impact of the project on the basis of the information provided and the observations made during the public hearing. 4) The competent authority assesses the environmental impact of the project. The final decision, together with a statement of reasons, is made available to the promoter and to the public (cf. Lambrechts 1996: 77-78).

But transposition was incomplete. First, more than one third of the Annex II projects were omitted. While Germany made 50 per cent of the Annex II projects

---

[84]   Interview with the EIA unit of the Bavarian Ministry of Environment (*Bayerisches Staatsministerium für Landesentwicklung und Umweltfragen*), Munich 09/98, and the EIA unit of the Federal Ministry of Environment (*Bundesministerium für Umwelt, Naturschutz und Reaktorsicherheit*), Bonn, 10/97.

[85]   *Gesetz zur Umsetzung der Richtlinie des Rates vom 27. Juni 1985 über die Umweltverträglichkeitsprüfung bei bestimmten öffentlichen und privaten Projekten, vom 20. Februar 1990, BGBl. I: 205.*

[86]   Changes became necessary in the *Abfallgesetz, Atomgesetz, Bundesimmissionsgesetz, Wasserhaushaltsgesetz, Bundesnaturschutzgesetz, Bundesfernstraßengesetz, Bundeswasserstraßengesetz, Bundesbahngesetz, Personenbeförderungsgesetz, Gesetz über den Bau und den Betrieb von Versuchsanlagen zur Erprobung von Techniken für den spurgeführten Verkehr, Luftverkehrsgesetz.*

subject to a mandatory EIA and imposed strict thresholds for some of the others, more than one third of the projects listed in Annex II of the Directive were not included in the *UVPG*. The formal exclusion of certain projects, which German policy-makers did not consider to become ever subject to an EIA or which were already subject to an EIA within media-specific authorization procedures, conforms to the German practice of limiting administrative discretion. It contradicts, however, the Directive, which requires member states to make decisions on a case-by-case basis or by using threshold criteria. Upon complaints of German environmental organizations, the Commission opened infringement proceedings for the incorrect transposition of the Directive in 1990, which resulted in a ruling of the ECJ upholding the position of the Commission.[87] The Court held that Germany had failed to meet its obligations by excluding entire classes of projects listed in Annex II from the requirement for environmental impact assessments. The *UVPG* also exempted projects from an EIA, if they had been publicly announced before its enactment in 1990 (§ 22, 1 UVPG). A Bavarian environmental organization, the *Bund Naturschutz in Bayern*, attacked this provision by appealing before the Bavarian Administrative Court against the authorization of the construction of two new sections of motorway in Bavaria in 1991, which had proceeded without EIA. The Bavarian Court referred the case to the European Court of Justice under the Art. 234 ECT preliminary ruling procedure (ex-Art. 177). In this case, the ECJ confirmed the direct effect of the EIA Directive, which Germany would have had to transpose by July 1988.[88]

Second, the federal ordinances, which were to specify the regulations on how to integrate the EIA into medium-specific authorization procedures, were delayed for several years. As a result, industrial plants became only subject a mandatory EIA in 1992, when the ordinance regulating the authorization procedures under the Federal Immission Control Act was finally amended.[89] But even then, the threshold values for industrial plants were so high, that the EIA has been applied to only a small number of cases (Staeck and Heinelt 2001: 66).

Third, public participation under the *UVPG* is restricted. § 9 (1) UVPG remits to the norm of the German Administrative Procedure Act (*Verwaltungsverfahrensgesetz/VwVfG*), which stipulates that only those 'whose interests are affected by the project' can express opposition (§ 73 (4) VwVfG). The European Directive, by contrast, provides for the participation of the 'public concerned' (Art. 2, 6, 9), not only of those directly affected by a project. Grounding the right of participation in the individualized harm, which a project might cause, leads to the exclusion of formally organized interests, which claim to represent the public interest. Moreover, public participation is reduced to the presentation of opposing views (*Ein-*

---

[87] C-1995/301, 22.10.1998.

[88] C-1992/396.

[89] *Neunte Verordnung zur Durchführung des Bundes-Immissionsschutzgesetzes (Verordnung über Genehmigungsverfahren, 9. BImSchV), vom 29. Mai 1992*, BGBl. I: 1001.

wendungen) to the project, which waters down the cooperative approach of the Directive since promoter and affected individuals are somehow assumed to have purely antagonistic interests regarding the project.

Finally, the acceleration and simplification legislation, which was adopted after German unification to promote economic reconstruction in the East but which was subsequently extended to the whole Federal Republic, reduced the number of projects subject to a compulsory EIA. The Investment Relief and Residential Building Land Act of 1993,[90] for instance, has allowed the *Länder* to waive the EIA in regional planning procedures. The Approval Procedure Accelerating Act of 1996 neither requires an EIA nor public involvement in the planning approval.[91] The acceleration and simplification laws also significantly curb the time period, in which the concerned public may voice concerns against a planned project (cf. Schink 1998).

In light of the rulings of the European Court of Justice and the revision of the EIA Directive in 1997, Germany felt compelled to amend its EIA legislation. The categories for which an EIA was compulsory were to be extended. The *Länder* had to make similar adaptations for those projects that fell into their responsibilities. More 'progressive' administrators both at the federal and regional level have favoured a fundamental reform of German environmental legislation by replacing the various media-specific laws through a uniform Environmental Code (*Umweltgesetzbuch*).[92] In 1997, an expert commission had presented a draft law accommodating the requirements of both the two EIA and the IPPC Directives. The general part was to provide the basis for integrating environmental protection in all relevant policy areas, and for public participation, information and access to the courts. It also contained rules for regulation in planning and licensing, environmental audit and assessment, and liability law. The draft proposal, however, never entered the legislative process due to the resistance of the sectoral administration (*Fachverwaltung*) of the *Länder*, which feared losses of competencies. The Federal Ministry of Justice also insisted that the *Bund* lacked certain competencies necessary to enact the Environmental Code (e.g. in the area of water resource management). German industry had reservations, too, since it feared more regulation. The red-green government of Chancellor Schröder, which came into office in 1998, had first committed itself to adopting an Environmental Code but then abandoned the idea resorting to an '*Artikelgesetz*' (framework law), which should ensure compliance

---

[90] *Gesetz zur Erleichterung von Investitionen und der Ausweitung und Bereitstellung von Wohnbauland vom 22. April 1993*, BGBl. I: 2141, see also *Verkehrswegeplanungsbeschleunigungsgesetz vom 16.12.1991*, BGBl. I: 2714 and the *Gesetz zur Vereinfachung der Planungsverfahren für Verkehrswege vom 17. Dezember 1993*, BGBl. I: 2123.

[91] *Gesetz zur Beschleunigung und Vereinfachung immissionsschutzrechtlicher Genehmigungsverfahren vom 12. September 1996 (BGBl. I: 1354)*. The new law provides for the acceleration and simplification of authorization procedures in the area of air pollution control, waste disposal, nuclear power, and water management.

[92] Interview with the EIA unit of the Federal Ministry of Environment, Bonn, 10/97.

with several EU environmental Directives including the IPPC Directive and the revised EIA Directive. But the *Artikelgesetz* has met similar resistance as the Environmental Code. Since Germany did not deliver, the Commission opened another infringement proceeding in 1999 for non-compliance with the 1998 ECJ ruling. The Commission referred the case to the ECJ in December 2000 and announced that it would ask for a daily penalty of EUR 237,600 if the ECJ convicted Germany. Since Germany had not implemented the revisions of the EIA either, the Commission opened a second infringement proceeding after the deadline for transposition had expired in March 1999. The case was also referred to the ECJ in November 2000.[93] Germany escaped a conviction in both cases since it finally passed the *Artikelgesetz* in July 2001.[94] But only one month after the Commission had withdrawn the two cases, it opened another infringement proceeding in December 2001 for the incorrect transposition of the EIA revision. It took Germany more than 14 years to correctly transpose the EIA Directive, and the problems appear to continue with its revision of 1997.

Practical application and enforcement has been as deficient as the formal implementation of the EIA Directive. First, by merely integrating the EIA into existing authorization procedures, the legal and administrative structures continue to be fragmented and counteract the cross-media approach of the Directive. As a result, EIA requirements become diluted in the authorization process (Erbguth 1991). Second, the administrative directive regulating the practical application of EIA in Germany only came in 1995, ten years after the EIA Directive had been passed.[95] In the absence of any guidelines, public authorities applied the EIA regulations in accordance with existing administrative practice.[96] Once enacted, the new administrative directive has further watered down the cross-media, interdisciplinary approach of the EIA Directive as well as the requirements for public participation. It has essentially reduced the practical application of EIA to the compliance with environmental standards, which exist for the different media. The licensing procedure (*Genehmigungsverfahren*) obliges the competent authority to approve a project, when legal standards are met. In the sector of air pollution control, for instance, the EIA merely requires conformance with air limit values specified in the *TA Luft*. Only if the unified consent procedure (*Planfeststellungsverfahren*) applies, public authorities have discretion in consent decisions. But even then licensing authorities have not sufficient training and expertise for doing integrated assessments. As a result, cross-media effects are not considered. Nor are alternatives to the project assessed (Mergner 1997). Public participation is also

---

[93] C-2000/408, 8.11.2000.

[94] *Gesetz zur Umsetzung der UVP Änderungsrichtlinie, der IVU-Richtlinie und weiterer EG-Richtlinien zum Umweltschutz vom 27.7.2001*, BGBl. I: 1950, 2.8.2001.

[95] *Verwaltungsvorschrift zur Ausführung des Gesetzes über die Umweltverträglichkeitsprüfung vom 18. September 1995, Gemeinsames Ministerialblatt 46 (32): 669.*

[96] Interview with the EIA unit of the Bavarian Ministry of Environment, Munich, 03/97.

often inadequate since it barely complies with the minimum legal requirements of the German legislation, such as the Administrative Procedure Act (*Verwaltungs-verfahrensgesetz*), according to which only affected parties are entitled to take part in the authorization procedure (see above). Moreover, public involvement takes place at a relatively late stage in the procedure, where major changes to planned projects are no longer considered possible or necessary. Competent authorities are usually reluctant to involve the public earlier, since they fear that it would delay the entire process. Finally, since the authorizing authority merely summarizes the overall environmental impacts of a project without preparing a separate EIA document, the public has no opportunity to voice its opinion on the outcome of the assessment (Commission of the European Communities 1993a: 4a).

All in all, changes to administrative practice have been marginal. The consequent application of the EIA regulations would not only presuppose additional staff-power and expertise at the subnational levels. The medium-specific structuring of German environmental policy-making does not allow for a cross-media assessment of environmental impacts. This would require the establishment of coordination mechanisms spanning across different authorities at different levels of government.

The ineffective application of the EIA procedure has provoked the opposition of domestic actors. Yet, restrictive legal provisions have often prevented environmental organizations from invoking the rights under the EIA Directive. Directly affected citizens can organize themselves in local citizen groups (*Bürgerinitiativen*), but their standing in administrative procedures and court appeals is limited by their scarce resources. Like in Spain, domestic actors have not sufficient staff-power, expertise, and money to systematically appeal against the deficient application of EIA regulations. The organization, which has most successfully influenced the implementation of the EIA Directive, is the EIA Association (*UVP Förderverein*), whose members include local authorities, consultancies, universities, local pressure groups, and environmental organizations. The EIA Association is at the centre of the policy community that emerged around the implementation of the EIA. It primarily aims at providing relevant information to experts, authorities and other interested parties in the EIA sector. The Association maintains contact with similar organizations in other member states. But it acts as an intermediator between the European Commission and German authorities rather than as a catalyst of pressures from above and from below on public actors to comply with the Directive. The EIA Association is careful not to intervene in particular procedures and refrains from criticizing final assessments. Nor does it engage in any verification activities (Staeck and Heinelt 2001: 65).

But even if environmental and citizen groups succeed in bringing legal action against public authorities, German jurisdiction tends to support the restrictive interpretation of the EIA followed by the public administration in the practical application of the Directive. The Federal Administrative Court (*Bundesverwaltungsgericht*), for instance, has causally linked the EIA procedure to the consent decision. It considers the absence of an EIA or procedural errors in the EIA procedure only as infringements of the *UVPG*, if those errors or the carrying out of an EIA would have led to a substantially different consent decision, that is, would have prevented the authorization of the project, altogether. The case-law of the highest administrative court in Germany, which is based on counterfactual reasoning putting the burden of proof on the plaintiff, contradicts the participatory approach to the EIA Directive. The EIA is to provide the competent authority with additional information from both the public concerned and other administrations and shall promote a more holistic perspective on the environmental impacts of a project, which may not necessarily result in a non-consent decision but the consideration of alternatives or corrective measures (cf. Heinelt et al. 2001). While pressure from below has been of limited effect, the European Commission has not exerted much pressure from above either. It appears to be reluctant to interfere in the practical application of the EIA Directive. Despite a series of complaints lodged by German environmentalists and citizen groups, the Commission has refrained so far from opening proceedings against issues that go beyond formal implementation.

In brief, while domestic mobilization has been quite effective in pushing Germany towards formal compliance with the EIA Directive, it has been far less successful in overcoming the resistance of public authorities to practically apply and enforce it.

Summing up, the EIA Directive has produced considerable policy misfit in Spain and Germany, which caused problems of compliance for both countries. Spanish and German environmentalists alike mobilized against the ineffective implementation of the EIA Directive. While Spanish environmental groups have been less successful in inducing the Commission to exert additional pressure for adaptation, several infringement proceedings forced Germany to remedy its incomplete transposition legislation. Domestic adaptational pressure for improving practical application and enforcement of the EIA procedure is as diffuse and ineffective as in Spain since local environmental and citizen groups in both countries have only limited resources.

Access to Information: Misfit in Spain and Germany

The Policy and its Development

The difficulties of monitoring and enforcing the growing number of European environmental policies motivated the Commission to take measures, which render information on the environment more easily available to the citizen. The most important policy in this area is the Directive on Freedom of Access to Information on the Environment (Access to Information Directive).[97] The Directive is based on the idea that broader access to environmental information (AI) increases transparency and openness thereby encouraging citizens to participate more actively in the pro-

---

[97]   90/313/EEC; OJ L 158, 23.6.90.

tection of the environment. Entailing elements of procedural and communicative regulation, the Directive aims at ensuring that the public has free access to information on the environment held by public authorities. Public authorities in possession of information on the environment, or bodies with public responsibility for the environment, have to make such information available to any natural or legal person upon him or her request without his or her having to prove a specific interest or concern. Public authorities must respond to a request for information within two months. They can ask a reasonable charge for providing the information. The Directive only allows for refusal if a request concerns information, which affects public security, the confidentiality of the proceedings of public authorities, international relations, national defence, matters which are under legal inquiry, commercial and industrial confidentiality, the confidentiality of personal data, material voluntarily supplied by third parties, unfinished documents and internal communications, and where disclosure of the information may damage the environment. Authorities can also deny a request when the request is manifestly unreasonable or formulated in too general a manner. The public must have the possibility to seek judicial review against the refusal of, or failure to provide requested information. Beside the obligation to make environmental information available upon request, member states are called upon to actively provide general information to the public. They were also asked to send a report on their experience with the Directive after four years to the Commission, which was then to submit a report to the Parliament and the Council.

The publication of the Commission's report on the working of the AI Directive and the proposal for its review had been scheduled for the end of 1998 but was delayed until June 2000 due to the late receipt of the member state reports and the signing of the Aarhus Convention,[98] the latter of which requires numerous changes to the Directive. Rather than revising the Directive, the Commission decided to adopt a new one.[99] The new Directive will replace Directive 90/313. It aims, first, at paving the way towards the ratification by the European Community of the Aarhus Convention through the alignment of the proposal to the relevant provisions of the Convention. Second, it seeks to adapt Directive 90/313 to developments in information and communication technologies reflecting thereby the changes in the way information is created, collected, stored and transmitted. Third, the new Directive will correct the shortcomings identified in the practical application of Directive 90/313. For example, a more comprehensive definition of environmental

---

[98] Convention on Access to Information, Public Participation in Decision-Making and Access to Justice in Environmental Matters, United Nations Economic Commission for Europe (UN/ECE), Fourth Ministerial Conference on Environment for Europe, Aarhus, 23.-25.6.1998, Conference Document ECE 7CEP/43/Add1./Rev., cf. Bugdahn 2001; Wates 1996.

[99] Amended Commission Proposal for a Directive of the European Parliament and of the Council on Public Access to Environmental Information, Document 501PC0303, 21.01.02.

information is provided, referring to human health and safety, cultural sites and built structures, and cost-benefit and other economic analyses. The scope of public authorities subject to the Directive is significantly broadened. They include government and all other national, regional or local public administration, which are no longer required to hold environmental responsibilities; it is sufficient that they are 'legal persons' being entrusted by law or other arrangements with the operation of services of general economic concern which affect or are likely to affect the state of elements of the environment. The scope of the exceptions for refusing information has been further clarified. Access to information may only be refused if disclosure of the information will adversely affect the interests protected by the exception. The public interest served by the disclosure of the information has to be weighed against the interests served by the exceptions. Access to the information shall be granted if the public interest served by disclosure outweighs the interest protected by an exception. The deadline is shortened to one month (instead of two in Directive 90/313) within which public authorities have to supply the requested information to applicants. Detailed provisions on charges, for which public authorities can ask for supplying the requested information, are also included. The supply of information cannot be made subject to the advance payment of a charge. The new Directive also makes reference to the EC Data Protection Directive[100] to ensure greater coherence between European data protection legislation and freedom of information. Finally, the member states are required to provide for a non-judicial complaint investigation procedure to allow for a quicker and less costly settlement of appeals cases than in the judicial review.

Like in case of environmental impact assessment, the Access to Information Directive originated at the European rather than at the member state level. In May 1984, the European Parliament adopted a resolution on 'the compulsory publication of information by the European Community'. The 4th Environmental Action Programme (1987-1992) mentioned the need for an EC Freedom of Information Act, which should oblige Community institutions and member states alike to provide access to information. The Commission committed itself to drafting a European Directive. While the European Parliament and environmental organizations welcomed the initiative, it received little support among the member states, for which the disclosure of environmental data was a highly sensitive issue. Only Denmark, France, Luxembourg, and the Netherlands, which already had their own regulations on the issue, favoured a European Directive. The UK and Germany, in particular, voiced strong opposition. In both countries, environmental data were handled as highly confidential. A European access to information policy would impose substantial institutional and legal adjustment costs. The British position, however, fundamentally changed when Chris Patten became Secretary of Environment and launched a reorganization of British environmental policy in 1988, including a turn around on the disclosure of environmental data. The Environ-

---

[100] 95/46/EC, OJ L 1995 281/31, 23.11.1995.

mental Protection Act of 1990, which was a response to growing international and domestic pressures on British regulatory practices, grants far-reaching public access to environmental information. The British government had a strong incentive to upload its public access policy to the European level in order to avoid competitive disadvantages for its industry. The new public registers give competing foreign firms access to data from which they could draw information on production processes (Héritier, Knill, and Mingers 1996: 237). The British change of mind left Germany isolated in the Council. Germany could no longer maintain its opposition without risking to lose its reputation as an environmental leader. Since the open and vague texture of the Directive left the member states considerable discretion concerning exemptions and the practical arrangements for providing the information (cf. Kimber 2000: 169-173), Germany hoped to be able to reduce adaptational costs by minimizing the effect of the European policy on its regulatory structures in implementation (Héritier, Knill, and Mingers 1996: 238).

Spain: Increasing Pull and Push

In Spain, access to information held by public authorities has traditionally been highly restricted and not freely available either to ordinary citizens or to non-governmental organisations (NGOs). During the Franco dictatorship, the Law of Administrative Procedure of 1958[101] allowed for restricted access to certain documents or records, provided the person could claim a personal, direct and legitimate interest. Public authorities were obliged to reply to all requests within six months. If a request was not answered within this time, the 'administrative silence' had to be understood as a refusal, against which an administrative appeal was to be lodged before the case could be taken to the administrative courts. In the democratic Constitution of 1978, Spain recognized the general right of citizens to access files and public registers, except where such access could affect national security and defence, criminal investigations or personal privacy (Art. 105b). The constitutional provision, however, has merely declaratory character since it does not constitute a subjective right to which citizens can refer in claiming access to a specific piece of information.

Several sectoral laws have obliged authorities to make public certain environmental information[102] but only in justified cases; they do not grant a general right to access to information. The Spanish Supreme Court somehow strengthened the right to access to information by broadening the interpretation of 'personal, direct and legitimate interest', which has enabled societal interest associations, working for the public benefit, as a whole to participate in administrative proceedings and to request information from public authorities (Sanchis Moreno 1996: 230).[103]

---

[101] *Ley de Procedimiento Administrativo de 17 de julio 1958.*

[102] For a list see De la Torre and Kimber 1997: Fn. 10.

[103] A court decision of 23 June 1992 granted a Spanish environmental organization (AEDE-NAT) access to information on nuclear waste storage facilities which had been refused

In brief, Spanish legal provisions and administrative practices of granting access to information only in justified cases was in sharp contrast with the AI Directive, which demands general access to information for anybody only to be refused in justified cases. This misfit has produced a high pressure for adaptation in the implementation of the AI Directive. The costs of such adaptation do not only lie with the additional working load for the administration, which has to supply the information. Broader access to information provides the public with an effective means of controlling administrative behaviour, including the monitoring of compliance with environmental legislation. It also allows for more transparency in administrative decision-making. Not being used to public scrutiny, Spanish administrators have shown little enthusiasm in an effective implementation of the AI Directive.

When the AI Directive had come into force in December 1992, the European Commission started receiving complaints from Spanish environmental organizations about the non-transposition of the Directive. The State Secretariat of Environment (national environmental authority) informed Spanish environmental organizations of the preparation of a draft law to 'publicize and provide access to information on the environment and to establish the criteria for collective legitimization'. Yet, this draft law, intended to properly implement the Directive, was never submitted to the Spanish Council of Ministers for approval (Sanchis Moreno 1996: 232). In March 1993, the Spanish government notified the Commission that the new Law on the Legal Regime of Public Administrations and Common Administrative Procedures passed in 1992 had implemented the AI Directive.[104] The law amends the general provisions on access to information established by the Law of Administrative Procedures of 1958 basically taking into account the constitutional precepts and the jurisprudence of the Supreme Court. Article 35 of the new law recognizes the right of the Spanish citizens to have access to documents in the course of administrative proceedings, in which he or she is involved. The right of access to registers and files held by public authorities is granted to 1) citizens who (individually or collectively) can claim a legal right or legitimate interest, and 2) associations and organizations representing economic or social interests.

But unlike claimed by the Spanish government, the regulations of the new law concerning access to information do not properly implement the AI Directive. First, the right of access to information is only granted to Spanish citizens and associations and not to any natural and legal person irrespective of nationality as stipulated by the Directive. Second, access is limited to files and documents which are held in administrative registers, are part of a record and belong to completed administrative proceedings; in order to get information relating to proceedings

---

by the Ministry for Industry and Energy. In another court decision, Greenpeace won access to information on the use of drift nets, which had been denied by the Ministry of Agriculture, Fisheries and Food.

[104] *Ley 30/1992 de 26 de noviembre, de Régimen Jurídico de las Administraciones Públicas y del Procedimiento Administrativo Común*, BOE N° 285, 27.11.1992.

which are unfinished and not completed, the requester needs to prove an interest which is in contrast to the provisions of general access in the Directive. Third, the law of 1992 fixes a time period of three months for the public administration to respond to requests whereas the Directive requires member states to respond as soon as possible and at the latest within two months. Fourth, the Spanish law includes exceptions, which go beyond those permitted by the Directive. These exceptions refer to files and documents, which form part of registered and closed files and dossiers. Fifth, the convention of 'administrative silence' is confirmed. If a request is not answered within three months, this is to be considered as a refusal. The Directive, however, states that public authorities have to justify a refusal.

Due to the incomplete transposition of the AI Directive, complaints to the Commission continued.[105] In March 1994, the Commission sent the Spanish government an Art. 169 Letter. At the same time, Spanish environmental organizations sought to invoke the 'direct effect' of the AI Directive (Sanchis Moreno 1996: 234-235), and the Commission also intervened in some cases of refusal of information. *Izquierda Unida*, a small party of the radical left, submitted an albeit non-legislative proposal before the Spanish Parliament for a comprehensive right to access to information, increasing the pressure on the Spanish government to correctly transpose the AI Directive (Sanchis Moreno 1996: 223). In view of the combined pressure from above and from below, the Spanish government prepared a new draft law on the Right of Access to Information on the Environment, which was enough for the Commission to stop infringement proceedings. In 1995, the AI Directive was finally transposed into national law.[106] The law explicitly states that its purpose is to complement already existing national legislation in order to comply with the AI Directive.

The new law closely follows the structure and content of the AI Directive. It even includes one positive amendment. If the information is available in different formats, the requester can choose the format in which he or she likes the information to be provided. Nevertheless, the new law does not fully implement the European norm. First, the right of access to information keeps being restricted – it only applies to nationals or residents of states forming part of the European Economic Area (EEA). Second, the principle of administrative silence is upheld. A request not answered within two months is to be considered as refusal, even if no justification is provided. Third, authorities can make charges for supplying requested information. There is no mentioning that such charges must not exceed a reasonable cost. Each authority is free to make its own charges[107] and there is no obligation to

---

[105] Two complaints were lodged by Greenpeace, one in February 1993, and the other in January 1994 (De la Torre and Kimber 1997).

[106] *Ley 38/1995 de 12 de diciembre, sobre el Derecho de Acceso a la Información en Materia de Medio Ambiente*, BOE N° 297, 13.12.1995.

[107] However, the 1989 Public Charges and Prices Act (*Ley 8/1989, de 13 de abril, reguladora de las Tasas y Precios Públicos*) is applicable, of which Art. 7 establishes a principle of equivalent charges.

inform the requester in advance of the charges that will be made. Finally, the law took on all the exceptions of the Directive without further specifying them, which leaves public authorities considerable scope for arbitrary interpretation (cf. Sanchis Moreno 1996; De la Torre and Kimber 1997).

Since transposition of the AI Directive was still incorrect, complaints of societal actors to the Commission continued. In February 1996, Greenpeace presented a third complaint to the Commission arguing that Spanish legislation still did not effectively implement the AI Directive. In September 1997, the Commission sent Spain two Reasoned Opinions, the first taking up the infringement proceeding opened in 1993 but suspended in 1994, and the second following up another proceeding for incorrect transposition, which the Commission had initiated in 1995. Both cases were referred to the ECJ in May 1999. Facing a negative ruling of the ECJ, the Spanish Ministry of Environment started to revise the transposing legislation addressing the concerns of the Commission, which was enough for the Commission to withdraw the two cases and to terminate the proceedings in July 2000 after Spain had enacted the revisions in January 2000.[108]

Like formal transposition, practical application has not been effective in Spain either. Most of the Spanish regions have not enacted any legislation to develop the national law on access to information on the environment. Murcia was the only Autonomous Community that issued some proper legislation.[109] Hence, there are no practical arrangements for regulating access to information. Many requests are simply not answered. Or, high charges are imposed.

A study of 105 cases of request for environmental information, collected by AEDENAT, Greenpeace, and the *Fons de Documentació i Medi Ambient*, illustrates the difficulties in gaining access to information. 80 per cent of these requests, all, however, prior to the formal transposition of the Directive in 1995, are cases of non-reply. Only in five per cent, access to information was provided. Against five cases, an administrative appeal was lodged. The administration decided in two cases, in both granting access to information. In five cases, a complaint was submitted to the Commission. The case study also reveals a change in administrative behaviour. Whereas in the past no charges had been made for providing information, some authorities started to take advantage of the possibility of charging provided by the AI Directive. The charges were sometimes so high that they could only be interpreted as a means of dissuading requests (Sanchis Moreno 1996).

A more recent study of the Catalan Ministry of Environment on the practical application of the AI Directive in Catalonia presents a more positive picture

---

[108] *Ley 55/1999, de 29 diciembre de 1999, de Medidas Fiscales, Administrativas y del Orden Social*, BOE Nº 312, 30.12.1999, whose Art. 81 amends *Ley 38/1995*.

[109] *Ley 1/1995, de 8 de marzo de 1995, sobre Protección del Medio Ambiente en la Región de Murcia, Título VI*. The regional legislation, which pre-empted central state transposition, remedies one major deficiency of the central state transposition law of 1995 as it defines 'administrative silence' (no reply within two months on whether a request for information is approved) as approval of the request instead of a denial.

(unpublished study, 5.2.1997). It acknowledges that there are no regulations specifying the application of the Directive, for instance regarding exceptions. But Catalan environmental authorities claim that this has not led to any problems. The only difficulty mentioned is the delimitation of aspects of confidentiality. Charges have been seldom made, and if so, only for photocopying. About 3600 requests for information are made each year which fall under the Directive (out of 30,000). The fact that 95 per cent of these requests were made by phone, indicates that most requests refer to 'trivial', that is, non-contentious issues and are made by ordinary citizens rather than societal organizations. Accordingly, the rate of response is high. Only one unit reports to have refused about 10 per cent of the requests invoking exceptions under the Directive. There was no case of appeal. At the same time, both Catalan environmentalists and consultancies complain about the difficulties in pursuing information from public authorities. Requests are not answered, 'fussy' excuses are given for not handing out information, or high fees are imposed.[110]

Environmental groups have mobilized against both the flawed transposition and the ineffective application of the AI Directive. Greenpeace has filed three complaints with the Commission for incorrect transposition, which resulted in the opening of two infringement proceedings. It also challenged in the Spanish High Court an exclusion by an authority of information on fisheries from the definition of information on the environment, and obtained a favourable judgement.[111] Two major Spanish environmental organizations, AEDENAT and CODA, launched nation-wide campaigns providing information brochures, which explain the rights of access to information including standardized forms by which information can be requested as well as appeals against refusal be made. Documentation centres collect cases of non-compliance with AI requirements (Sanchis Moreno 1996). At the local level, however, public demand for environmental information is still low. Local environmentalists and citizen groups have only limited resources to push their rights to information against the resistance of public administration. The implementation of the AI Directive reveals the same dilemma as for the EIA Directive. AI has triggered some mobilization, which, however, has emerged from actors that already had the capacity to mobilize, in this case (trans)national environmental organizations. Actors with less resources have hardly been empowered by the new right to access to information since the Directive is not effectively applied and they are too weak to pull the policy down to the local level. On the broader societal level, the misfit between the EU policy and national practices effectively constraints the mobilizing effect of the Directive. Suing public authorities takes about three years and involves high costs, also because citizens need legal assistance. Hence, there is little systematic pressure from below on public authorities to correctly apply the AI Directive. Moreover, unlike transposition, the Commission has been little responsive to complaints about the flawed application of the Directive.

---

[110] Interviews with DEPANA, Taller d'Enginyeries, and KPMG, Barcelona 03/97.

[111] *Sentencia 106/91 de la Audiencia Nacional de 30 de junio de 1993.*

Instead of taking action, it referred most complainants to national appeals procedures (Bugdahn 2001: 64). Pressure from below has helped to improve formal implementation. But it has only had limited effect on practical application.

Germany: Increasing Pull and Push

There is no comprehensive right to (environmental) information under German law. Art. 5 of the German Constitution only grants the right 'to inform oneself without restraint from all *sources of information that are open to everybody*' (emphasis added). It does not entail the access to administrative records, which are conceived of as for internal purposes only. The Administrative Procedure Law (*Verwaltungsverfahrensgesetz/VwVfG*), which provides the legal basis for access to information in administrative proceedings, specifies the conditions under which information is provided to the public (Erichsen 1992). Its provisions are based on the principle of 'restricted access to records', which grants the public access to information only in justified cases, that is in the context of certain administrative procedures and *when third parties can claim a 'legitimate' interest*. Like in case of the EIA, individuals have to prove that their subjective rights may be affected by an imminent administrative decision to get access to official files.[112] The final stages of the administrative decision process are in any case protected from disclosure. Decision drafts and preparatory work on the formulation of the decision are withheld from the public, both before and after the administrative procedure is completed (§ 29 VwVfG). In other words, access to information is an essential element in the legal protection of substantive rights of the individual but not designed to guarantee the participation in public decision-making (Winter 1996: 81). This approach, based on the confidentiality of information in possession of public authorities, clearly contradicts the Directive, which aims to enhance transparency and public participation in administrative decision-making by granting a comprehensive right to access to environmental information irrespective of interest and procedural context.

In light of this significant misfit, German authorities perceived the implementation of the AI Directive as extremely costly. Resistance against predominantly cognitive costs for having to change the dominant problem-solving approach came in the disguise of arguments of administrative efficiency. Administrators claimed that citizens would use their right to access to information in an 'offensive' or 'abusive' matter, for instance, by mobilizing against the authorization of major projects (*sic*). Given the inclination of German citizens to engage in legal disputes (*Streitkultur*), public authorities expected to be flooded by information requests. Broad access to information would impede the smooth operations of public administrations that were forced to answer numerous requests instead of working

---

[112] Even then, it is not sufficient that a person claims that she was not given access to information required to defend her legal interests; she has to demonstrate that having had access to the information would have led to a different outcome (§ 46 VwVfG).

towards improving environmental protection (Schink 1994: 192). The additional work load for collecting, organizing and providing the requested information would require the hiring of additional personnel, which was unfeasible under the increasingly tight budget constraints (Turiaux 1994: 23-23).

German industry was equally opposed to the new Directive. It feared that broad access rights would considerably delay the authorization of projects and provide competitors with confidential information about production processes (see EIA).

Given the broad resistance, the transposition of the Directive was seriously delayed. When the deadline of 31 December 1992 had expired, the German government had not more than a draft bill, which was highly contested among the various ministries concerned. Several drafts prepared by the BMU had already failed due to the opposition of the *Länder*[113] as well as some federal ministries, which had insisted on a more restricted interpretation of public authorities that are subject to the obligation of the AI Directive to provide access to environmental information.[114]

Since transposition was delayed, German environmental groups filed several complaints with the Commission. When the Commission opened an infringement proceeding in 1993 and sent a Reasoned Opinion in 1994, Germany finally enacted the Environmental Information Act (*Umweltinformationsgesetz/UIG*).[115] The new Act, however, did not correctly transpose the Directive. In some instances, the formulations of the *UIG* were vague and ambiguous, which left ample space for interpretations against the 'spirit' or even the wording of the Directive (Scherzberg 1994; Schwanenflügel 1993). In other instances, the *UIG* adopted a restrictive interpretation of the Directive. First, the definition of public authorities subject to the AI regulations was far more limited than in the Directive. Only those authorities

---

[113] *Empfehlungen der Ausschüsse zum Entwurf eines Gesetzes zur Umsetzung der Richtlinie 90/313/EWG des Rates vom 7. Juni 1990 über den freien Zugang zu Informationen über die Umwelt, BR-Drs. 797/1/93 vom 7.12.1993* and the *Stellungnahme des Bundesrates vom 17.12.1994 zum Gesetzesentwurf der Bundesregierung, Nr. 2 und 3 zu Art. 1, § 3 1 UIG-Vorschlag, BT-Drs. 12/7138 vom 23.3.1994.*

[114] Among them were the Ministries of the Interior, of Transport, of Agriculture and of Finance. The issue was finally settled by the environmental committee of the *Bundestag*, which unanimously decided that all authorities which did not have their main responsibility in the area of environmental protection but which were subject to environmental regulation, such as the Ministry of Transport or Agriculture, or subject to the AI Directive (*Bundestag-Drs. 12/7582 vom 18.5.1994*). Yet, after the *UIG* had been adopted, the Ministry of Transport issued an administrative directive which stated that it did not consider road planning authorities to be subject to the *UIG* (Letter of the Federal Minister of Transport of 18.8.1994, *AZ: StB 15/2363 Vma 94*). It took the intervention of the European Commission and a clarifying letter of the German Chancellor to persuade the Ministry to abandon its strategy of non-compliance (Gebers 1996: 106).

[115] *Gesetz zur Umsetzung der Richtlinie 90/313/EWG des Rates vom 7.6.1990 über den freien Zugang zu Information über die Umwelt vom 8. Juli 1994*, BGBl. I: 1490, 15.7.1994.

primarily involved in environmental protection matters were required to provide information to the public, excluding transport, zoning and public works authorities (Art. 1, §§ 2 (2), 3 (1) UIG). The Directive, however, also includes those authorities 'possessing information' (Art. 2b; cf. Scherzberg 1994: 735).

Second, the exemptions were not clearly defined. The categories such as privacy, the protection of public safety, firm secrets and intellectual property were rather broad (§§ 7, 8 UIG). What constituted a firm secret was left to the firm's own evaluation (§ 8 (2), cf. Engel 1992: 112). The provision of a list of issues that would not qualify as commercial or business secrets, was considered but finally dropped in the legislation process (Bugdahn 2001: 202-203). Likewise, there were no criteria according to which public authorities might declare an legislative or authorization procedure as ongoing and hence exempt information collected in the course of the open procedure from the AI requirements (§ 7 (1) Nr. 2 UIG). The interpretation of preliminary proceedings, which were also excluded from free access to information, was so broad as to include all administrative proceedings thereby allowing German authorities to continue their practice of not-disclosing draft decisions or files related to their preparation (§ 7 (1) Nr. 2 UIG). According to the Directive, these documents should be open for access to everyone. The German government justified this broad interpretation of the Directive on the grounds that authorities must not be disturbed in their opinion formation and decision-making by information requests (Meyer-Rutz 1993). Finally, it was left unclear whether the implementation of regulations were considered part of the norm-setting process, since the *UIG* made an exemption for authorities involved in legislation and judiciary (§ 3 (1) Nr. 1 UIG; cf. Scherzberg 1994: 735; Winter 1996: 88-90; SRU 1996).

Third, the *UIG* did not specify the form in which authorities had to provide access. Contrary to German administrative practice, the competent authority had a choice of how to supply requested information. Thus § 4 (1) UIG left it to the discretion of the administration whether to grant direct access to official records (*Akteneinsicht*) as previewed by the Directive.[116]

Finally, there were no clear criteria for charges which public authorities could impose for the provision of information. While the AI Directive allows member states to 'make charges for supplying information', which 'must not exceed reasonable costs' (Art. 5), German provisions required that 'charges should cover the prospective costs' (§ 10 (1) UIG). The Ordinance regulating the fees that could be charged for providing environmental information left authorities wide discretion.[117] While the provision of 'simple' oral or written information was to be free of charge, cases that entailed an unusual amount of work (e.g. when numerous references had to be deleted in order to protect the interests of public or private parties) could be charged between EUR 1022 and EUR 5122. Even the denial of a request could cost up to EUR 1000 (Scherzberg 1994: 744). The Federal Minister of Envi-

---

[116] For a more extensive evaluation see Engel 1992; Gebers 1993; Scherzberg 1994; SRU, Sachverständigen Rat für Umweltfragen 1996; Winter 1996: 106.

[117] *Umweltinformationsgebührenverordnung*, BGBl. I: 3732, 7.12.1994.

ronment justified this regulation as a means to prevent abuse and inflationary requests for information.[118] High charges have been encouraged by the Finance Ministries of *Bund* and *Länder*, which interpreted the AI Directive as an opportunity to increase revenues (Lenschow 1997: 38). The Budgetary Committee of the German Parliament had also objected when the Environment Committee sought to make the provision for charges more access-friendly, arguing that the budget constraints did not allow for more generous rules (Bugdahn 2001: 201).

Access to Information was further legally restricted by new legislation aimed at facilitating investments in the infrastructure of the New *Länder*, but which were subsequently extended to the rest of Germany (OECD 2001: 30; see EIA section). The acceleration and simplification laws allow for the exclusion of private objections, of suspension activities due to objections and of lawsuits for quashing administrative decisions as well as for the fixing of time limits for official statements of views and inquiries (Erbguth 1999: 18).

At the *Länder* level, the implementation response was mixed (cf. Bugdahn 2001). As the transposition of the AI Directive got delayed, all 16 *Länder* issued administrative directives for a transitional application of the AI Directive due to the high number of public requests claiming direct effect of the AI Directive. Yet, administrative directives are externally non-binding. Moreover, most of the *Länder* regulations closely – sometimes even literally[119] – followed the federal draft legislation and, hence, were as evasive and vague as the later *UIG* allowing for far-reaching exemptions and high charges to discourage public requests (Gebers 1996: 98-99). Some *Länder* started to draft their own legislation but withdrew when the *UIG* came into force since it has the status of concurrent legislation not leaving any scope for regional transposition responses. Some *Länder*, however, passed general Freedom of Information Acts. Brandenburg, Berlin, Schleswig-Holstein, and North-Rhine-Westphalia grant their citizens broader access to files and documents than the *UIG*, not only in the area of the environment.[120]

In light of the incorrect transposition, the Commission continued to receive complaints from German environmental groups and citizens. In March 1994, three months before the *UIG* was finally adopted, the Commission had already notified the German ambassador to the EU that it considered the German definition of 'authorities with environmental responsibilities' as well as the provisions on charges in the German draft bill as a violation of the Directive (Gebers 1996: 101). Since

---

[118] Klaus Töpfer, *Bundestags-Plenarprotokoll 12/220, Verhandlungen des Deutschen Bundestages, 12. Wahlperiode, Stenographische Berichte*, 15.4.1994: 19081.

[119] Compare the *Referentenentwurf* (informal *UIG* draft) of the federal government of December 2, 1992 with the *Rundschreiben* (circular) of the Bavarian Ministry of Environment of January 10, 1993 (*Info-Dienst des Bund Naturschutz in Bayern, Nr. 129, Juli 1993*), or the *Runderlass* (circular) of the Environmental Ministry of North-Rhine-Westphalia of January, 28, 1993 (*Ministerialblatt für das Land Nordrhein-Westfalen, Nr. 20, 22.3.1993*).

[120] For a link to the legal texts see http://www.datenschutz.de/(de)/recht/gesetze.

Germany refused to rectify the provisions accordingly, the Commission opened an infringement proceeding after the *UIG* had been enacted. It accused Germany for the failure to correctly transpose the AI Directive with regard to four points: 1) failure to specify provisions for a differential treatment of information requests separating out information on items to be withheld rather than denying access altogether, 2) failure to restrict the exemption clause for ongoing procedures, 3) failure to contain the charges for the supply of information to a non-excessive level, and 4) failure to prevent authorities from charging the refusal of a request (Kimber 2000). Although the Federal Ministry of Environment was aware of the ineffective transposition of the Directive, the German government denied that its transposing legislation was incorrect. Since Germany did not respond to the Reasoned Opinion sent in September 1996, the Commission referred the case to the ECJ in June 1997 where it was decided in 1999.[121] In its ruling, the ECJ upheld the position of the Commission. It found that Germany had failed to provide access to information during administrative proceedings where the public authorities had received information in the course of those proceedings,[122] to specify in the *UIG* that information to be supplied in part where it was possible to separate out confidential information and to prescribe that a charge was to be made only where information was in fact supplied. It did not follow the Commission on the failure to specify that charges must be reasonable thereby accepting the argument of the German government that other provisions and general principles of German administrative law could ensure charges not to exceed a reasonable amount. Since the Commission had merely based its case on the incorrect transposition of the AI Directive, the ECJ refused to consider whether its application led in practice to excessive, unreasonable charges. The court provided, however, an interpretation of what constituted 'reasonable costs' rejecting any amount that could dissuade a person from seeking to obtain information. Also, the ECJ found that Germany had transposed the Directive incorrectly by allowing authorities to charge the refusal of a request.

In the meantime, a new coalition of Social Democrats and the Green Party came into power in September 1998, which promised a general widening of citizenship rights. The red-green coalition even declared its intention to introduce a general Freedom of Information Act.[123] Yet, a substantial policy change has not

---

[121] C-1997/217, 9.9.1999.

[122] The ECJ had already clarified this point in its preliminary ruling on a case, which was initiated by two major German environmental organizations, BUND and NABU, and which the High Administrative Court of Schleswig-Holstein referred to the ECJ under the Art. 234 (ex-Art. 177) procedure. The ECJ ruled that preliminary investigation procedures related 'solely to proceedings of a judicial or quasi-judicial nature, or at least proceedings which would inevitably lead to the imposition of a fine if the offence, whether administrative or criminal, was established' (C-1996/321, 17.6.1998, ECR. I – 3809, para 29).

[123] *Aufbruch und Erneuerung – Deutschlands Weg ins 21. Jahrhundert. Koalitionsvereinbarung zwischen der Sozialdemokratischen Partei Deutschlands und Bündnis 90/DIE GRÜNEN, Bonn, 20. Oktober 1998*: Section 9, No. 13.

emerged from the change in government. Germany finally signed the Aarhus Convention, which will grant German environmental and citizen groups legal standing (*Verbandsklagerecht*). But the German government attached a declaration that the legislative consequences of the Convention would have to be carefully considered before it could become binding under international law. Moreover, Germany made clear that the implementation of the Convention must not counteract efforts towards deregulation and speeding up procedures. The new government made no progress on the revision of the *UIG* either. After the conviction of the ECJ for incorrect transposition of the Directive, which came one year after the first 'green' Environmental Minister had come into office, environmental organizations increased their pressure on the German government (Bugdahn 2001: 211). In November 2000, the German government finally published a draft law, which provided a 'slim' revision of the *UIG*.[124] The proposed *Artikelgesetz* (framework law) linked the changes necessary to bring Germany in compliance with the AI Directive to the transposition of the IPPC Directive and the revision of the EIA Directive. Since the new law was delayed, the Commission opened another infringement proceeding under Art. 228 for non-compliance with the ECJ ruling of 1999.[125] A Reasoned Opinion was sent in March 2001. In July 2001, the *Artikelgesetz* was finally passed (see EIA section). The new law accommodates the main concerns of the Commission, which closed the infringement proceeding in October 2001. Art. 21 of the *Artikelgesetz* provides for a differential treatment of information and limits the exemptions of ongoing procedures. Moreover, access to information has to be granted or refused within two months; simply communicating the intention to provide the information is not sufficient. Art. 22 lowers the maximum charges from EUR 5112 to EUR 500 and facilitates the waiving of fees. Charges can only be made for the actual supply – and not for the refusal – of information.

Whether these changes will have an impact on the practical application of the Directive remains to be seen. So far, administrative practice has been as flawed as the legal implementation (OECD 2001: 30). The vagueness of German regulations has allowed public authorities to interpret them against the 'spirit' or even the wording of the Directive. When the Directive came into force at the beginning of 1993, German authorities were confronted with extensive requests from the public for information concerning the environment (Gebers 1996: 97). Some public interest groups systematically tested the practical application of the Directive by asking different authorities for the same type of information. In 1993, for example, the *Öko-Institut* requested public authorities in the *Land* Hesse for a list of all plants that were required to make a declaration of emissions (Küppers 1994). The *Öko-Institut* has also collected other cases concerning violations of the Directive, published newspaper articles on the topic, and made several complaints to the Euro-

---

[124] *BR-Drs. 674/00, Gesetzentwurf der Bundesregierung. Entwurf für ein Gesetz zur Umsetzung der UVP-Änderungsrichtlinie, der IVU-Richtlinie und weitere EG-Richtlinien zum Umweltschutz vom 10.1.00.*

[125] C-1997/217, 9.9.1999.

pean Commission. The *Bund für Umwelt und Naturschutz Deutschland* (BUND) launched a test campaign in 1995 asking regional environmental offices in the *Land* North-Rhine-Westphalia for particular information (Küppers 1995). In 1994, the BUND had already sent a written request to the Environmental Ministries of the *Länder* for a list of plants that had to inform the public about the safety measures taken. The information was meant to help monitoring compliance with the German Ordinance on Accidents, which implements the Seveso Directive (Lenius 1994). Likewise, the BUND contacted 26 local authorities requesting information on the implementation of measures for the cleaning-up of groundwater (Lenius and Ekhardt 1995). The environmental magazine *Öko-Test* commissioned a study in 1995 to test the application of the *UIG* in all the 445 districts of Germany (*kreisfreie Städte und Landkreise*) asking for data on drinking water quality and contaminated sites (Becker 1995).

The various studies disclosed similar deficiencies in the applications of the AI Directive. Like in Spain, the most frequent response to information requests was administrative silence. If public authorities responded, they often denied the request arguing that they had no responsibility for the environment, that the requested information was not concerning the environment, or that the information was protected by law. Even where civil servants were in principle willing to grant access to information, they felt uncertain about violations of regulations for data protection and commercial secrets, as a result of which they tended to interpret the exceptions of the *UIG* in a restrictive manner. In other cases, public authorities refused to provide information which they considered part of an open administrative procedure, including the licensing of industrial plants or judicial proceedings. While not all authorities charged for the same type of information, fees frequently amounted to several EUR 100 to discourage the systematic collection of data. This has particularly been the case when sensitive 'enforcement information' has been involved. Requesters, for instance, have been asked to pay for the costs of making public officials familiar with the German legislation (Kimber 2000: 186). Or they have been charged for the working time a civil servant had to spend on compiling the information, which could result in a bill of EUR 100 for making three copies. This 'deterrence' strategy has proven to be quite effective; citizens tend to withdraw their request when being informed about high fees.[126]

In short, German environmental groups have strongly mobilized against the ineffective application of the AI regulations. Like their Spanish counterparts, they organized information campaigns, issued publications, launched several series of systematic requests for information, collected and documented cases of refusal, lodged administrative appeals, filed law suits, and sent complaints to the Commission. Domestic mobilization has been stronger than in Spain, given the superior resources of German environmental organizations and citizen groups. But its effect has been largely restricted to improvements in formal compliance. Appealing against the wrongful application of the AI Directive may not only take up to six

---

[126] *Landesbeauftragter für Datenschutz Schleswig-Holstein, Annual Report 1997*: 69.

years before an administrative court ultimately decides the case. The impact of a positive judgement is limited since it is confined to isolated cases. Moreover, German courts have often adopted an equally restrictive interpretation of the AI legislation as public authorities, up to the point that their jurisdiction has been characterized as 'environmental information hostile' (Röger 1997: 885; cf. Partsch 1998). Several courts denied the Directive direct effect when its transposition was delayed.[127] Or they did not properly check the applicability of exemptions[128] and confirmed that information obtained in an ongoing administrative procedure was to be exempted from access.[129] Also, lower administrative courts have sought to avoid giving a judgement at all by referring the case back to the administration for further investigation and discussion.[130] The European Commission has not been too responsive to German complaints about the inadequate application of the AI Directive either. Only once, in 1996, the Commission opened infringement proceedings against Germany for public authorities in the *Land* Schleswig-Holstein only provisionally responded within the mandatory two month period, and failed to supply the requested information.[131] The case was referred to the ECJ in February 2000[132] but the Commission withdrew in November 2001 after the *Artikelgesetz* had been passed. In light of the restrictive application of the AI provisions by public authorities, as well as the restrictive jurisdiction of administrative courts, citizens and environmental groups have preferred to invoke their rights under the AI Directive only as a last resort to get access to information they could not obtain otherwise. First of all, they exploit the information rights granted by licensing procedures or ask local and regional members of parliament to make a request.

Like the EIA Directive, the AI Directive is a clear example for an EU policy causing serious problems of compliance for environmental leaders and laggards alike. Spanish and German authorities have resorted to similar strategies in trying to circumvent the requirements of the Directive. So did Spanish and German environmental organizations in their attempt to pull the AI Directive down to the domestic level. While their repeated complaints to the European Commission induced pressure from above, which has pushed both countries closer to formal compliance,

---

[127] *Verwaltungsgericht Stade, 7. Kammer Lüneburg, Urteil vom 21.4.1993, 7A 79/92*; cf. Wegener 1994.

[128] *Oberverwaltungsgericht Münster, Urteil vom 27.9.1993, Az 21 A 2565/92.*

[129] *Verwaltungsgericht Gelsenkirchen, Urteil vom 1.9.1994, 8 K 6370/93.*

[130] *Verwaltungsgericht München, Urteil vom 26.9.1995, M 16 K 93; Verwaltungsgericht Freiburg, Urteil vom 8.11.1996, 9 K 1341/95; Verwaltungsgerichtshof Mannheim, Urteil vom 10.6.1998, 10 S 58/97.*

[131] Report from the Commission to the Council and the European Parliament on the experience gained in the application of Council Directive 90/313/EEC of June 7 1990, on Freedom of Access to Information on the Environment, COM (2000) 400: 7.

[132] C-2000/029, 1.2.2000.

domestic mobilization has remained rather ineffective in generating pressure from below on practical application and enforcement because it is too diffuse to bring about any systematic improvements.

Eco-Audit Management: Misfit in Spain and Germany

The Policy and its Development

The 4th Environmental Action Programme stressed the need to develop more integrated policies, which entail a multi-media approach to the environment and aim at cutting pollution at the source. The 5th Environmental Action Programme (1993-2000) singles out public and private enterprises as being among the key targets of, and participants in, environmental policy. The Community Eco-Management and Audit Scheme (EMAS), established by a Regulation in 1993,[133] seeks to combine the two goals. It aims at the evaluation and improvement of the environmental performance of industrial activities and the provision of relevant information to the public. While the environmental impact assessment applies only to proposed projects, the related concept of environmental auditing is intended to minimize effects on the environment when an installation is already operating. EMAS is a voluntary instrument, which should provide incentives for enterprises to introduce an environmental management system for assessing and improving industrial activities and providing the public with adequate information. Thus, EMAS relies on self-regulation, communication and market incentives rather than on command-and-control instruments to fight environmental pollution. By introducing an environmental management system, enterprises shall find more efficient management solutions reducing economic costs. They may also gain a competitive advantage in the market, if consumer acknowledge better environmental performance in their product choices. Finally, the implementation of EMAS may ease public monitoring since information on environmental performance is regularly provided.

The Regulation requires the member states to establish a voluntary scheme which would recognize industrial sites where firms had set up, within the framework of a company policy, environmental management systems, including regular audits and public reports on environmental performance. An enterprise, which enters the scheme, has to carry out an environmental review of all environmental issues and impacts associated with activities at the site. On the basis of this analysis, it will adopt an environmental policy that defines the environmental objectives as well as the environmental programme for the site and the management scheme to achieve these objectives. Regular environmental audits, at a frequency of between one and three years, will ensure effective implementation. The company has to prepare an annual environmental statement, designed for public consumption, which includes *inter alia* a description of the company's activities on the site con-

---

[133] 1836/93/EEC; OJ L 168, 10.7.1993.

cerned, an assessment of all the major environmental problems related to these activities, as well as a summary of the numeric data on polluting emissions, waste production, consumption of raw materials, energy and water, and noise. The environmental statement, together with a presentation of the environmental policy, programme, management system and audit procedure, is submitted for review and validation by an independent accredited environmental verifier. The environmental audits and declarations are subject to public scrutiny. The auditor has to validate the environmental statement of an enterprise, which is made available to the public. Only then can the enterprise register its site with the competent public authority. In its annexes, the Regulation sets out in more detail requirements of the environmental policy, programme and management systems as well as the auditing, the accreditation and functioning of the auditors, and the information to be provided to the competent body at the time of registration. Having direct effect, the EMAS Regulation does not require transposition into national law. However, the member states have twelve months after the entry into force of the Regulation to establish a system of registration for sites implementing EMAS, as well as of accreditation for independent environmental inspectors and the supervision of their activities.

In December 1998, the Commission presented a proposal for the revision of the EMAS Regulation, which aimed at making it more compatible with the standard for environmental management systems created by the International Standardization Organization (ISO 14001) in 1996.[134] The revision was formally adopted in February 2001.[135] Sites registered under EMAS can now automatically receive ISO certification. Moreover, the principle of site based audits is replaced by the principle of organization wide application, which is followed by the ISO standard. The scope of EMAS was extended to other sectors of the economy, not just manufacturing sites, and to the public administration. The new Regulation also includes provisions for greater public access to the environmental statements, a stronger legal compliance clause, better oversight of the environmental verifiers as well as more specific criteria for continuous improvement in environmental performance.

Like the Access to Information Directive, the EMAS Regulation originated in the United States, where 'compliance audits' were introduced in the 1970s due to a growing number of major accidents at industrial sites, more stringent environmental regulations and liability cases. In the late 1980s, the International Chamber of Commerce suggested a standardization of environmental audit procedures at the international level (cf. Waskow 1997: 1-4). The European Commission took up the initiative (Malek and Töller 2001: 43). It supported the idea of market-oriented, voluntary instruments. Moreover, a Europe-wide scheme would prevent the proliferation of national standards, such as the British Standard BS 7750, which could lead to distortions within the Common Market. But unlike already existing schemes, the initial proposal of the Commission favoured a compulsory scheme

---

[134] For a comparison between EMAS and ISO 14001 see Malek and Töller 2001: 48-50.

[135] 761/2001/EC; OJ L 114/761, 24.4.2001.

with uniform and binding standards for certain sectors of industry and for installations above a certain size. The voluntary approach was only adopted after industry and some member states had expressed their opposition (Haigh 2001: 11.8-3).

Diverging national preferences introduced further changes to the Commission proposal (Waskow 1997: 6-9). The UK was a strong advocate of a Europe-wide voluntary eco-management scheme. It anticipated low adjustment costs since the voluntary and procedural character of such a scheme would conform to the British regulatory tradition. Moreover, the UK had piloted a national environmental management scheme in 1992, the British Standard BS 7750, which provided a model for the European Regulation granting the UK a 'first mover' advantage, also with regard to contested consultancy market. Finally, the British government saw EMAS as a way to harmonize the far-reaching data publication obligations, which the Environmental Pollution Act imposed on British industry (Héritier, Knill, and Mingers 1996: 255-256). The reasons, which made Britain a strong supporter of the Regulation, turned Germany into its major opponent. In fact, Germany was the only member state that opposed the Regulation till the very end of the negotiations. The German Ministry of Environment rejected the voluntary character of EMAS and the absence of any substantive regulations, which left the member states too much discretion in formulating the requirements for EMAS and would therefore undermine the high level of environmental regulation in Germany. Resistance also came from the Ministry of Economics and German industry. Both were concerned that EMAS could bring more rather than less regulation at the national level. German firms also argued that EMAS would result in a competitive disadvantage for them since registered sites were required to continuously improve their environmental performance *irrespective* of their starting level. Furthermore, the provisions of the publication of environmental statements were seen as a threat to commercial confidentiality (Waskow 1997: 6). Despite the voluntary character of EMAS, German industry felt that firms would have to register if they were concerned about their public reputation. Moreover, insurance companies and banks could exert pressure on companies to carry out environmental audits to reduce environmental risks (Héritier, Knill, and Mingers 1996: 258). German environmental organizations equally rejected EMAS since it neither contained objective assessment criteria nor real obligations for industry to provide information to the public (Malek and Töller 2001: 45).

Despite strong domestic opposition, the German government ultimately gave its consent in the Council. After the ratification of the Maastricht Treaty, the Regulation could be adopted by qualified majority instead of unanimity. Germany decided to use its veto power as long as it was still possible to negotiate some concessions. Thus, it successfully pushed for a provision in the Regulation on the application of the best available technology as a benchmark by which continuous improvement in environmental performance is measured.[136] The obligation to comply with existing

---

[136] The BAT requirement, however, was qualified by the 'economically viable application of the Best Available Technology' (Art. 3a), which has nowhere been legally defined.

environmental legislation added another element of substantive regulation (Lübbe-Wolff 1994: 47). The EMAS Regulation was finally adopted in June 1993, after a relatively short period of negotiations compared to the EIA and the AI Directives.

Spain: Neither Pull nor Push

Environmental management and audit schemes were not entirely new to Spain. Like the UK and Ireland, Spain developed a national management system based on the British Standard 7750. While EMAS is far more demanding than the Spanish UNE scheme, pressure for adaptation has been limited due to the voluntary character of the European scheme.

Spain fully incorporated the required system of registration, accreditation and supervision into its already existing 'infrastructure for the quality and safety of industry' for ensuring certain quality and safety standards applied to industrial activities.[137] Existing regulations were modified by a Royal Decree in 1995, which makes reference to EMAS.[138] The EMAS Regulation itself was officially implemented shortly after.[139] While sites implementing EMAS have to register with the environmental authorities, accreditation and supervision of the environmental verifiers is handled by the national accreditation society (*Entidad Nacional de Acreditación/ENAC*). *ENAC* is a private non-profit organization, which authorizes private and public actors (inspectors, laboratories, consultancies) to verify and certify the compliance of enterprises with industrial quality and safety standards. Its governing body comprise representatives of the central and regional administration as well as of 'interested parties' (industry, trade unions, environmentalists) and is assisted by various expert committees. The requirements, which the environmental inspectors or auditors have to fulfil in order to be accredited with *ENAC*, are set by the Spanish Association for Standardization (*Asociación Española de Normalización/AENOR*).[140] Like *ENAC*, *AENOR* is a private, non-profit organization of pluralist composition, which sets standards for the normalization and certification of industrial activity (UNE), including EMAS.[141] The competent authority for registration at the central state level is the Ministry of Environment. After having approved a site for registration, the Ministry of Environment asks the Ministry of Industry and Energy to include the site in the Register of Industrial Sites (*Registro de Establecimientos Industriales*). The national system of accreditation and super-

---

Given the vagueness of the concept, it is doubtful whether this can be considered a substantive requirement for EMAS.

[137] *Ley 21/1992, de 19 de julio, de Industria, titulo III.*

[138] *Real Decreto 2200/1995*, de 28 de deciembre, BOE N° 32, 6.2.1996.

[139] *Real Decreto 85/1996*, de 26 de enero, BOE N° 45, 21.2.1996.

[140] UNE 66500, which corresponds to European Norm EN 45000.

[141] UNE 77/801/94.

vision serves for those Autonomous Communities, which do not wish to establish their own system. They have the right to designate proper accreditation bodies, which have to comply, however, with the requirements of the Royal Decree implementing EMAS at the national level. Regional bodies must also seek approval by the national Council for the Coordination of Industrial Security (*Consejo de Coordinación de la Seguridad Industrial*) of the Ministry of Industry and Energy. Catalonia was the first region to establish its own system.[142] Even before the Catalan government had set up its own accreditation unit, it denied inspectors accredited by *ENAC* the right to do audits in Catalonia, which resulted in a conflict over competencies before the Constitutional Court.

All in all, the implementation of EMAS gave only rise to marginal legal and administrative changes both at the national and the regional level. One can argue about the extent to which EMAS fits the Spanish legal and administrative structures. The effective implementation of EMAS would have required some legal and administrative changes, which go beyond the mere absorption into the existing structures. There are, for instance, no clear criteria for the accreditation and supervision of the auditors. But Spanish administration has felt no need to engage in further changes. Spanish industry has shown hardly any interest in pulling EMAS down to the domestic level. So far, only few companies have registered under EMAS (85 sites in 2001, which accounts for about two per cent of all sites registered in the EU so far; Haigh 2001: 11.8-5). Most of them are multinational companies. For small and medium sized enterprises, which account for 90 per cent of the Spanish business sector, the costs of implementing an eco-audit and management scheme are high, and so are their concerns to reveal their degree of (non)compliance with environmental legislation. Moreover, companies are often unclear how the international, the national and the European audit schemes relate to each other. They tend to aim for the international ISO 14001, which was introduced in Spain in 1996 and which is less demanding and better known.[143]

In the absence of any domestic mobilization in favour of EMAS, public authorities have little incentive to make the implementation of EMAS more effective. The Spanish State and some of the Autonomous Communities offer companies public subsidies to help financing the investments necessary to implement environmental audit and management schemes. Yet, if Spanish firms apply, they prefer to use the funds for the less demanding ISO 14001, which is equally accepted by insurance companies and banks as well as the interested public. While registered

---

[142] *Decreto 230/1993, de 6 de septiembre*, DOGC N° 1806, 8.10.1993; cf. *Orden de 17 de agosto* 1993, DOGC N° 1809, 15.10.1993; *Decreto 115/1996 de 2 de abril*, DOGC N° 2192, 10.4.1996.

[143] Interview with the EMAS unit of the Catalan Ministry of Environment, Barcelona, 03/97, and the Spanish Business Association (CEOE), Madrid, 03/97.

EMAS sites have been held out a prospect of facilitations in obtaining permits and licences,[144] the necessary regulations have not materialized so far.

Germany: Pull Without Push

The German position towards EMAS was very hostile in the first place (see above). The emphasis of the Regulation on industrial self-regulation, procedural rules and voluntary participation does not fit the German interventionist, command-and-control approach in environmental policy, which is based on uniform and compulsory rules and substantive technological requirements. Like in the cases of EIA and AI, the cognitive misfit motivated German resistance against EMAS. While EMAS found little acceptance among German administrators, even German industry appeared to prefer the interventionist problem-solving approach for its legal predictability and clarity (Héritier, Knill, and Mingers 1996: 69-70). Moreover, like public administration, industry considered the provisions for making data on the environmental performance of companies public as a breach of the principle of commercial and industrial confidentiality. Finally, German companies feared that EMAS would lead to even more administrative regulations and procedures (permitting, inspection, auditing, verification, registration) instead of decreasing regulatory burdens.

Next to cognitive costs, the creation of accreditation and certification bodies required a series of organizational changes. Unlike the UK or Spain, Germany did not have any national provisions for an environmental audit system, into which EMAS could have been integrated. It had to either establish new competent authorities or to allocate additional responsibilities to already existing units. The German Ministry of Environment wanted to absorb EMAS as far as possible into the existing regulatory structures. Like most of the *Länder*, it had a clear preference for a state-oriented concept where a public authority, such as the Federal Environmental Office (*Umweltbundesamt*), would be responsible for the accreditation and supervision of the environmental verifiers. Only technical assistance should be assigned to the Association of Accreditation (*Trägergemeinschaft für Akkreditierung*), an organization representing the interests of German industry (Héritier, Knill, and Mingers 1996: 260).

The Ministry of Economics and German industry, by contrast, pushed for a self-administrative, decentralized accreditation system under the responsibility of the industrial and trade associations (cf. Waskow 1997: 109-118). Once the European Regulation had been adopted, German industry abandoned its opposition and

---

[144] See the information brochures on EMAS distributed by national and regional ministries of environment: *Ministerio de Medio Ambiente: El Sistema Comunitario de Gestión y Auditoría Medioambientales en España*; *Consejería de Medio Ambiente y Desarrollo Regional de la Comunidad de Madrid: El Sistema Comunitario de Gestión y Auditoría Medioambientales en la Comunidad de Madrid*; *Departament de Medi Ambient de la Generalitat de Catalunya: El Sistema Ecogestión y Ecoauditoría de la Unión Europea en Catalunya*.

sought to use EMAS as an instrument for deregulation in the field of monitoring and licensing industrial plants. Implementing EMAS should allow to 'substitute' state control by voluntary self-regulation granting firms participating in the environmental audit scheme permitting facilitations and exemptions from certain monitoring requirements. By linking EMAS to the general debate on 'slimming the state' (*Schlanker Staat*) in order to prevent competitive disadvantages for German industry (*Standort Deutschland sichern*),[145] German industry succeeded in pressing for an implementation of the Eco-Audit Regulation which was closer to a market oriented approach of voluntary self-regulation than to the traditional German approach of binding state regulation as it was favoured by the environmental administration.

The implementation of EMAS was indeed a compromise between state interventionism and industrial self-regulation, which was reached after two years of tough and lengthy negotiations between the Federal Ministry of Environment and environmental groups, on the one hand, and major business associations, on the other (cf. Schneider 1995). Since the law enacting the EMAS Regulation (*Umweltauditgesetz/UAG*)[146] only came in December 1995, more than six months after the deadline had expired, some interim rules had been introduced to avoid competitive disadvantages for German companies and environmental verifiers (Malek et al. 2001: 108).

The *UAG* assigns accreditation authority to the German Accreditation and Admission Association for Environmental Verifiers Limited (*Deutsche Akkreditierungs- und Zulassungsgesellschaft für Umweltgutachter mbh/DAZU*). The *DAZU* is a 'joint venture' between industry and trade associations.[147] It officially accredits EMAS verifiers and supervises their activities. The *DAZU* is subject to the legal supervision of the Federal Ministry of Environment and its work is regulated by the Environmental Verifier Committee (*Umweltgutachteraus-*

---

[145] See for instance the recommendations of the *Sachverständigenrat 'Schlanker Staat'* (http://www.bundesregierung.de/inland/ministerien/innen/sachver00.html) or the December 1994 Report of the *Unabhängige Expertenkommission zur Vereinfachung und Beschleunigung von Genehmigungsverfahren (Schlichter Kommission)*. The *BMU* presented the new Environmental Audit Law (*UAG*) as 'an entry in the deregulation of environmental administrative law' (*BMU Pressemitteilung, Bonn, 13.12.1995: 1*), and the Federal Minister of Environment, Angela Merkel, assigned EMAS a 'pilot function' following the principle 'slimming the state - slashing bureaucracy' (*Umwelt Nr. 6/1995*). For a comprehensive overview of the deregulation debate in Germany and the role of EMAS in this context, see (Waskow 1997: 24-44).

[146] *Gesetz zur Ausführung der Verordnung (EWG) Nr. 1836/93 des Rates vom 29. Juni 1993 über die freiwillige Beteiligung gewerblicher Unternehmen an einem Gemeinschaftssystem für das Umweltmanagement und die Umweltbetriebsprüfung vom 7. Dezember 1995, BGBl. I: 1591.*

[147] *Bundesverband der Deutschen Industrie (BDI), Deutscher Industrie- und Handelstag (DHI), Zentralverband des Deutschen Handwerks (ZDH), Bundesverband Freier Berufe (BFB).*

schuß/UGA). The *UGA* is a pluralistic body consisting of representatives of the *Bund* and *Länder* administration, industry, trade unions, verifiers and environmental groups. It prepares general guidelines, which the *DAZU* has to apply in the accreditation and supervision of environmental verifiers.[148] Like the *DAZU*, the *UGA* is subject to legal supervision by the Environmental Ministry. A committee installed by the Federal Ministry decides on appeals against decisions taken by the *DAZU* on the accreditation and supervision of environmental verifiers. Audited companies have to register with the Chambers for Industry and Commerce (*Industrie- und Handelskammern*) or the Chambers for Crafts (*Handwerkskammern*).

In short, the mobilization of German industry in the implementation of EMAS led to a mixed system of industrial self-regulation and moderate state intervention, which went far beyond the legal and administrative changes required by the Regulation. The delegation of state powers for the setting of professional criteria and guidelines to a pluralistic expert committee is a 'novum in German legislation'.[149] While the delegation of state functions to private bodies is known in Germany, professional accreditation (*Berufszulassung*) used to be a matter of public authority. The enactment of administrative guidelines by a self-administrating body is not new to German administrative practice either. But regulating the activities of environmental verifiers is an intervention in the constitutional right of freedom of profession (Art. 12, Basic Law) and has to be based on state law. The composition of the Environmental Verifier Committee is somehow unique, too, since it comprises representatives of industry, trade unions, environmentalists, verifiers and public administration (cf. Waskow 1997: 119-130). Finally, the approach of using EMAS as an instrument of substituting state control through industrial self-regulation contradicts the German regulatory approach of command-and-control regulation, which explains the virulent opposition of the specialized administration at all levels against EMAS.[150]

The German environmental audit and management scheme correctly implements the EMAS Regulation. It has been argued that the system is seriously biased towards industrial interests, which have substantial control over the verification and certification process through the Accreditation and Admission Association (*DAZU*). Since the *DAZU* had been established as part of the interim rules until the EMAS Law (*UAG*) came into force, there were concerns that at least the first

---

[148] *UAG-Zulassungsverfahrensverordnung, BGBl. I: 2013, 18.12.1995*, regulating the accreditation and certification procedure; *UAG-Beleihungsverordnung, BGBl. I.: 2013, 18.12.1995*, establishing accreditation and certification guidelines for environmental verifiers.

[149] *Pressemitteilung des Bundesumweltministeriums vom 13.12.1995: 1.*

[150] For a contending view see the work of Christoph Knill, who argues that EMAS does not contradict the general core of German administrative traditions. First, EMAS would only supplement existing regulatory instruments rather than replace them. Second, industrial self-regulation would correspond to corporatist elements of German state tradition with its intermediary organizations assuming public functions (Knill 2001: 160-161).

group of accredited verifiers would have too heavy an 'industrial background' (SRU 1996: 100). Moreover, business interests are over-represented in the Environmental Verifier Committee compared to environmentalists (ibid.), although a coalition of public administration and societal organizations still holds a majority. Finally, the registration of EMAS sites falls under the responsibility of industry and trade associations. Despite the strong role of industry in the implementation of EMAS, there is little evidence for 'agency capture'. The work of the Environmental Verifier Committee, for instance, reflects a focus on substantive issues rather than disputes between the different interests represented (Malek et al. 2001: 109). While industry and trade associations have discretion in assessing whether a site complies with the EMAS requirements, environmental authorities have the final say, particularly when it comes to compliance with environmental law.

Its industry-friendly implementation has made EMAS a great success among German business. Germany accounts for 75 per cent of registered sites in the EU (OECD 2001: 116-117). EMAS is particularly popular among the chemical industry (Malek et al. 2001: 110-112). The high level of participation is not only explained by image concerns or more favourable insurance and credit terms. Industry has had clear expectations about the reduction of inspections and compliance controls by public authorities. Self-regulation should at least partly allow for the substitution of state control. Moreover, it was hoped that EMAS would lead to some deregulation in terms of a reduction of legal requirements, for instance, in licensing procedures. But the initiative of the Bavarian government to use EMAS as a 'means of 'slimming bureaucracy' was not able to win a majority among the other *Länder*. While deregulation attempts have largely failed, the concept of substitution has found its way into several federal and *Länder* regulations. Substitution follows the idea of 'functional equivalency', in which participating firms would be relieved of certain legal obligations if a roughly equivalent obligation existed in EMAS. For instance, EMAS documentation fulfils the reporting requirements of recycling and waste management laws. Bavaria has been a forerunner in promoting substitution. In an 'environmental pact' (*Umweltpakt Bayern*) of 1995, the Bavarian government agreed with Bavarian business and trade associations to increase participation in EMAS, reduce waste, improve energy efficiency, and increase the use of rail transport. In return, EMAS sites would be subject to lower reporting and documentation requirements, reduced inspections and fastened licensing procedures. Moreover, the Bavarian government promised to grant subsidies for environmental technology and EMAS participation (Böhm-Amtmann 1997). Other *Länder* have followed the Bavarian example. Yet, the impact has been rather modest so far. Industry complains that substitution does not go far enough to really reduce regulatory burdens (Malek et al. 2001: 116). Local authorities are still reluctant to replace state control by industrial self-regulation because they do not trust the environmental commitment of business interests.[151] In the revision of the EMAS

---

[151] Interview with the air pollution control unit of the Bavarian Ministry of Environment, Munich, 03/97.

Regulation, German business pushed for a provision that would 'privilege' companies registered under EMAS in licensing procedures. The draft law for amending the German EMAS law in light of the EMAS revision, however, does not contain such a privileging cause (*Öko-Audit-Privilegierung*).[152] It had originally been included in the draft of the *Artikelgesetz* intended to transpose the IPPC Directive and to accommodate the revisions of the EIA and AI Directives (see above). Yet, environmental organizations have successfully mobilized against such an 'opening clause' for EMAS.

All in all, the emphasis on compliance control and functional equivalency in German legislation and administrative practice has somewhat counteracted the principle of industrial self-regulation in the legal implementation of EMAS. The way EMAS has been applied makes it look like yet another regulatory instrument in the German command-and-control tradition with its focus on the application of BAT and a strict governmental oversight. The disappointment of German industry about the limited regulatory relief actually achieved by EMAS shows in the rapid growth rate of registrations under ISO 14001, which, by now, has outstripped that of EMAS (Kollman 2001).

To conclude, EMAS did not fit either the Spanish or German interventionist regulatory approach. While the voluntary character of EMAS does not give rise to compliance problems as such, it illustrates the importance of domestic mobilization. In Spain, industry showed little enthusiasm for EMAS as a result of which the policy was merely absorbed into existing regulatory structures. In Germany, however, industry pulled the policy down to the domestic level pressing for an 'industry-friendly' implementation, which resulted in substantial legal and administrative changes and the highest level of industry 'compliance' in the EU – despite the initial policy misfit.

---

[152] *Entwurf für ein Gesetz zur Änderung des Umweltauditgesetzes vom 16.1.2002*, http://www.bmu.de/download/dateien/uag_entw.pdf.

Chapter Five

Conclusions

The concluding chapter summarizes the major findings of the book in light of the theoretical propositions of the Pull-and-Push Model. The comparative case studies challenge conventional approaches to non-compliance with European Environmental Law, which cannot account for the observed variations between Spain, an alleged southern laggard, and Germany, one of the northern leaders. Nor can these approaches explain differing degrees of compliance with policies within each of the two member states. Then, the scope of the Pull-and-Push Model as an alternative explanation of member state compliance is discussed. The model is general enough to study compliance with European Law across different policy areas as well as with other forms of law beyond the nation state. The chapter concludes by considering the implications of the theoretical argument and the empirical findings for European environmental policy-making. The Pull-and-Push Model would expect the forthcoming enlargement of the European Union to exacerbate existing compliance problems with European Environmental Law. Like the Southern European member states, the Central and Eastern European candidate countries struggle with high policy misfit, limited implementation capacities, and low levels of societal mobilization in the adoption of the environmental *acquis*. In order to ensure its effectiveness and legitimacy, European Environmental Law has to be more responsive to the distinct environmental problems and economic concerns of its environmental late-comers.

Challenging the 'Southern Problem'

This book set out to challenge the commonly held view that non-compliance with European Environmental Law is mainly a 'Southern Problem'. The literature has identified several deficiencies in the political and social institutions of Southern European countries that are believed to impair their capacity to effectively implement and comply with EU environmental Directives. First, a fragmented, re-active and party-dominated policy process impedes the enactment of effective environmental regulations. Second, fragmented administrative structures and insufficient administrative capacities prevent the successful implementation and enforcement of environmental regulations. Third, a lower level of socio-economic development prioritizes consumerist over 'post-modern' values resulting in low environmental awareness. Finally, a lack of 'civic culture' gives rise to corruption, clientelism, and non-compliant behaviour.

The book found little evidence for the existence of a 'Southern Problem'. The analysis in chapter two showed that non-compliance with European Law has neither significantly increased over time, nor is it confined to the four Southern European countries. While there are no data on the absolute level of non-compliance, we can compare non-compliance across time and member states. If we control for the growth in European legal acts and in member states, the number of infringements has remained rather stable over the years. And while infringement data indicate that non-compliance is more pronounced in the South than in the North of the European Union, there is significant variation among the Southern European member states as well as across the North-South divide. Italy and Greece consistently top the list of the compliance laggards. But Spain and Portugal range closer to the EU average when it comes to the later stages of the proceedings. The opposite is true for Belgium and Germany, whose performance subsequently deteriorates and which quickly join the group of compliance laggards.

Chapter three challenged the 'Southern Problem' on theoretical grounds pointing to four explanatory problems of the approach. First, the southern member states are as diverse in their political and social institutions as they are with regard to their compliance patterns. Administrative structures are more fragmented in Spain and Italy, with their highly decentralized territorial systems, than in Greece and Portugal, which are both unitary states. Italy is the fourth-largest economy in Europe and its administration commands more resources than the administrations of Portugal and Greece, which belong to the poorest member states. Italy and Greece range much higher on corruption indexes than Spain and Portugal. Environmental awareness, finally, is higher in Greece and Italy than in Portugal and Spain.

Second, problems of administrative fragmentation, scarce resources, powerful veto players, corruption, clientelism, and weak support for environmental regulations are not confined to Southern European countries. German and Spanish regions have repeatedly blocked and delayed the legal implementation and practical application of EU environmental policies. Unlike the German *Länder*, the Spanish Autonomous Communities have no formal veto power in the legal implementation of EU Directives. But they hold key responsibilities in practical application, monitoring, and enforcement. While administrative capacity is higher in Germany than in Spain, German administrators appear to be incapable of adapting their belief-systems to more integrated problem-solving approaches and policy instruments, which challenge the traditional German approach of substantive, command-and-control regulation for the different media. The French and Belgian ratings on the corruption indexes, finally, are close to Spain and Portugal, and their levels of environmental awareness are as low as in the southern member states.

Third, national differences in the major explanatory factors specified by the 'Southern Problem' literature do not match the observed levels of non-compliance in the member states. For the three (northern) compliance leaders, we find a common pattern. Although they differ significantly with regard to the level of institutional fragmentation, Denmark, the Netherlands, and the UK share a high level of socio-economic development, solid administrative capacities, low levels of corruption, and relatively high levels of environmental awareness. The four top compliance laggards, by contrast, hardly fit the pattern. Italy, Belgium, Greece, and Portugal, are rather diverse in their levels of institutional fragmentation, administrative capacity, socio-economic development, corruption, and environmental awareness.

Fourth, non-compliance with European Environmental Law not only varies across countries but also across policies. Member states have less difficulty in complying with European air pollution control regulations than with policies on nature protection or access to environmental information. Germany is a case in point. Despite its reputation as an environmental leader, Germany's performance in the implementation of the Fauna, Flora and Habitat Directive resembles more the one of an environmental laggard (OECD 2001: 89-104). The same is true for compliance with more innovative environmental policies, such as the Environmental Impact Assessment Directive or the Access to Information Directive. Since the 'Southern Problem' literature focuses on country variables to explain implementation failure and non-compliance, it is unable to account for cross-policy variations within individual member states. While some studies try to combine country and policy variables, they hardly specify how the different factors interact.

The second part of chapter three developed a model that seeks to overcome the explanatory weaknesses of the 'Southern Problem' approach. While building on factors, which the literature has identified as relevant in explaining non-compliance, the 'Pull-and-Push' Model systematically links them in a way that they are able to account for both cross-country and cross-policy variation. The model rests on two assumptions. First, non-compliance arises only, if European policies incur significant costs. European policies that do not fit the regulatory structures of a member state cause significant costs of adaptation in implementation and compliance. Policy-makers have to change legal and administrative institutions, (re)allocate resources to invest in additional staff-power, expertise, and monitoring technologies, and fend off political opposition by regulated parties. The greater the incompatibilities between European and domestic problem-solving approaches, policy instruments, and policy standards, the higher the adaptational costs, and the less inclined are policy-makers and administrators to effectively implement and comply with the 'misfitting' policy. If, by contrast, a European policy is compatible with the domestic regulatory structure, problems of implementation failure and non-compliance are unlikely to arise.

Second, policy misfit is only the necessary condition for ineffective implementation and non-compliance. Pressure from domestic and European actors can significantly increase the costs of non-compliance. Citizens, environmental organizations, interest groups, political parties and the media may mobilize pulling the 'misfitting' policy down to the domestic level. Through media and information campaigns, lobbying and monitoring activities, administrative appeals and court litigation, domestic actors generate substantial pressure on policy-makers and public authorities to improve implementation and compliance with 'misfitting' European policies. Such pressure from below is particularly effective if domestic actors are able to mobilize the Commission to open infringement proceedings exerting pressure from above. Infringement proceedings tend to increase both the political and the material costs of non-compliance by publicly shaming member states and by imposing financial penalties, respectively. If member states get sandwiched between combined pressure from the domestic and the European level, the chances of costly policies to be effectively implemented and complied with increase considerably.

The 'Pull-and-Push' Model allows to reformulate the 'Southern Problem' in a less reductionist and more dynamic way. Southern European member states are more likely to run into compliance problems, because, first, they encounter greater policy misfit, and second, their policy-makers and administrators face less pressure from below that pulls 'misfitting' policies down to the domestic level. The industrial and environmental late-comers of the South are policy takers rather than policy makers. They have neither the policies nor the capacity to upload them to the European level. Consequently, European Environmental Law has been shaped by the competing interests of the northern first-comers that seek to harmonize their environmental policies at the European level. They differ in their regulatory institutions but share common environmental problems and economic concerns, which tend to be different from those of the southern late-comers. While the first-comers may drag along the late-comers up to their level of environmental protection, the 'leader-laggard' dynamics tend to produce European environmental policies that are not only costly for the southern late-comers to implement but often do not address their most pressing environmental problems either. High policy misfit undermines both the capacity and the willingness of the Southern European member states to effectively implement and comply with European Environmental Law.

Southern late-comers not only face greater policy misfit. Their closed political opportunity structures and lower level of socio-economic development discourage the domestic mobilization that is necessary to generate sufficient pressure from below and from above to increase the costs of non-compliance for domestic policy-makers, administrators, and regulated parties. Authoritarian statism and the weak institutionalization of environmental politics seriously constrain the access of environmental interests to the policy process. Moreover, their organization capacities are still relatively weak. Citizens and environmental groups often lack sufficient funding, staff-power, and expertise to effectively lobby for compliance and litigate against non-compliance with European environmental policies. While we find environmental mobilization in southern countries, it tends to be localized. Loosely structured networks of local environmental groups which mobilize against environmentally damaging activities in their neighbourhoods do often not generate sufficient pressure on national policy-makers and administrators to make them face the compliance costs of 'misfitting' policies. Nor do they necessarily turn to European institutions for assistance. Finally, the strong public concern for economic growth and the creation of employment limits the possibilities of environmental interests to mobilize the media and the public at large. Even though environmental awareness has been rising, the problems that enjoy issue salience are not necessarily addressed by European environmental policies that need to be pulled down to the domestic level. In sum, non-compliance is more prevalent in the South than in the North, because southern late-comers have more limited compliance capacities but face higher compliance costs and lower compliance pressure than the northern first-comers.

The comparative study on the implementation of six different European environmental policies in chapter four systematically tested the propositions of the 'Pull-and-Push' Model. The six policies were selected as to allow to systematically vary the explanatory factors of the model. While the Drinking Water Directive, the Environmental Impact Assessment (EIA) Directive, the Access to Environmental Information (AI) Directive, and the Eco-Audit (EMAS) Regulation have caused significant policy misfit for Spain and Germany, the two Air Pollution Control Directives were perfectly compatible with the German regulatory structures and consequently did not give rise to any compliance problems. Spain, by contrast, had to adapt its emission values and build-up adequate monitoring networks, which imposed significant costs, material and political.

For the ten misfit cases, the empirical findings confirm that adaptational costs create compliance problems for environmental leaders and laggards alike (see figure 5). The EMAS Regulation and the EIA, AI and Drinking Water Directives have required Germany and Spain to adapt their domestic problem-solving approaches, policy instruments, and/or policy standards. So did the two Air Pollution Control Directives in Spain. The two countries have equally resisted the legal and administrative changes necessary to correctly transpose and effectively apply the costly policies. For Germany, the compliance costs have been more of a cognitive than a material nature. The integrated and procedural approach of the EMAS Regulation, the EIA Directive and the AI Directive and their emphasis on public information and participation strongly contradict the German administrative tradition of substantive command-and-control regulation enacted for the different environmental media. The effective implementation of these 'New Policy Instruments' (see below) not only requires Germany to profoundly modify its highly fragmented legislation and administrative structures. German administrators have to change their belief-systems, too. In Spain, by contrast, regulatory traditions are less deeply entrenched. While the Spanish state tradition is equally hierarchical and secretive, the major challenge for Spain has been to build-up new regulatory structures in order to meet European requirements. Networks for monitoring air and water quality, for instance, were only weakly developed and measurement technologies were not up to standard. New administrative procedures had to be established and additional personnel and expertise were needed. Moreover, Spanish policy-makers and public authorities have had to balance the enforcement of environmental standards against strong political demands for employment and economic growth.

Despite the adaptational costs incurred, not all misfit cases resulted in non-compliance. Compliance improved when domestic actors mobilized and pressured public authorities and policy-makers to face compliance costs. Domestic actors seeking to pull European policies down to the domestic level usually mobilized the Commission to push the policy from above by opening infringement proceedings. The Drinking Water Directive in Germany is the only case where the Commission took legal action without receiving massive complaints of citizens and environmental organizations (*push without pull*). The infringement proceedings resulted in a condemnation by the European Court of Justice, which forced Germany to give-up its more than ten years of resistance against formally complying with the Directive. In the absence of domestic mobilization, however, it is questionable whether German authorities always properly apply and enforce the European drinking water quality standards.

In case of the Large Combustion Plant in Spain, the Commission remained inactive. But local mobilization was sufficient to pressure public authorities and regulated parties towards compliance (*pull without push*). Practical application of the Large Combustion Plant improved when citizen groups, supported by (trans)national environmental organizations, protested against the environmental damage associated with the emissions of one of Spain's (and Europe's) most polluting power stations. In response to massive public pressure, the management of the plant voluntarily committed itself to emission reductions that went well beyond the obligations under the Directive. While Spain succeeded in reducing its emissions considerably over the past years, the BAT requirements are still not systematically applied and enforced, particularly when it comes to small and medium-sized enterprises.

The EMAS Regulation is another case of pull without push. Given the voluntary nature of the Regulation, the Commission had little grounds to interfere in the implementation process. The mobilization of German industry in favour of the Eco-Audit Regulation brought about the legal and administrative changes necessary to ensure a high level of compliance, despite the misfit between the self-regulatory approach of the Regulation and German state regulation. German industry hoped that certification under EMAS would alleviate some of its regulatory burdens imposed by (state monitoring of) environmental regulations. While Germany accounts by far for the highest number of registered sites, the enthusiasm of German companies appears to diminish since EMAS has not resulted in any substantial deregulation as yet.

The Environmental Impact Assessment and the Access to Information Directives provide four of the ten misfit cases were both the Commission and domestic actors have become active and jointly exercised pressure on German and Spanish authorities to face compliance costs (*push and pull*). In both countries, environmental groups launched information campaigns, systematically tested their new rights under the Directives, collected and documented cases of non-compliance, sought administrative and judicial appeals, and submitted several complaints to the Commission. The Commission responded to domestic mobilization by opening infringement proceedings, two against Spain and Germany each for non-compliance with the AI Directive, and five against each of the two countries for violating the EIA Directive and its revision. The combined pressure from below and above pulled and pushed (albeit slowly) Spain and Germany towards legal compliance with the two Directives. Practical application has been improving, too. Unlike in the cases of the Drinking Water Directive in Germany and the Large Combustion Plant Directive in Spain, domestic mobilization has time and again challenged the restrictive administrative practice in both countries. It remains to be seen, however, whether pressure from below will sustain and be sufficient to ensure the effective application and enforcement of the two 'New Policy Instruments' so that they can help improve the implementation of other environmental policies (see below).

In the three cases where there was neither 'pull' nor 'push', compliance with 'misfitting' policies has remained low (*neither pull nor push*). Spain has not transposed the Industrial Plant Directive and the BAT requirement lacks legal substance, nor is it systematically applied and enforced. The Drinking Water Directive was legally implemented but monitoring has been insufficient as a result of which it is difficult to assess whether Spain is actually complying with European drinking water standards. The EMAS Regulation, finally, was fully incorporated into the Spanish legal system. While European requirements have been largely met, the effective application of EMAS would require some additional legal and administrative changes. But as long as Spanish industry shows little enthusiasm for EMAS, these deficiencies are unlikely to be remedied. The same is true for the Drinking Water and the Large Combustion Plant Directives. Spanish citizens and environmental groups are less concerned with water quality standards as long as the regions still fight over issues of water quantity. Likewise, public support is difficult to mobilize for technical issues, such as the application of BAT in air pollution control, particularly if investments into new abatement technologies appear to conflict with demands for more employment and economic growth.

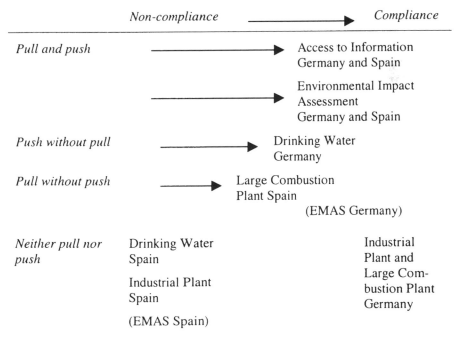

Figure 5   The Pull-and-Push Model and compliance with European Environmental Law

The findings of the 12 case studies confirm the expectations of the Pull-and-Push Model. In the two cases where policy misfit is absent (Air Pollution Control in Germany), we find full compliance. The opposite is true for the three misfit cases where there was neither pull nor push (Drinking Water, Industrial Plant, and EMAS in Spain). For the remaining seven cases, we find emerging compliance, which is most pronounced when German and Spanish authorities face joint pressure from below and above (pull and push).

Explaining Compliance with Law Beyond the Nation State: Pressure from Below and Above

The Pull-and-Push Model identifies two major variables in explaining the differing degree of compliance with European Environmental Law: policy misfit causing substantial compliance costs, and the mobilization of domestic and European actors increasing the costs of non-compliance. The causal relevance of both factors is not confined to the area of environmental policy. Several studies have shown that the EU's social, health, telecommunications, and transport policies can cause significant policy misfit for its member states resulting in serious implementation problems due to the adaptational costs involved (Caporaso and Jupille 2001, Kurzer 2001, Schneider 2001). The logic of policy misfit also holds outside the realm of regulatory policy, as works on European citizenship norms or refugee policy demonstrate (Checkel 2001b; Lavenex, 2001). In many cases the mobilization of societal actors proves to be crucial in overcoming the resistance of policy-makers and public authorities against the necessary legal and administrative changes. In the UK, for instance, women's organizations successfully pulled the Equal Pay and Equal Treatment Directives down to the domestic level, while the European Commission and the European Court of Justice pushed from above. The joint pressure from below and above generated the necessary changes in British legislation to bring the UK into compliance with EU Gender Equality Law (Caporaso and Jupille 2001).

The Pull-and-Push Model resonates well with compliance research in the International Relations literature (cf. Börzel 2002a; Börzel and Risse 2002). Irrespective of the theories they adhere to, most scholars agree that international norms and rules are most likely to be complied with if they are compatible with the domestic context including regulatory standards, political and administrative structures, problem-solving approaches, and collectively shared identities (Keohane 1984: 99; Breitmeier and Wolf 1995: 347-348; Underdal 1998: 12; Risse and Ropp 1999: 271; cf. Cortell and Davis 1996; Checkel 1997; Ulbert 1997). The more inconvenient international norms and rules are, the higher the compliance costs and the more likely are compliance problems to emerge. Moreover, several compliance studies have identified societal mobilization and international pressures as crucial factors in inducing compliance with 'misfitting' international norms and rules. The human rights literature, for instance, emphasizes the role of transnational non-governmental organizations (NGOs), which mobilize pressure on states from above, through coalitions with international organizations, and from below, through coalitions with domestic actors (Brysk 1993; Sikkink 1993; Keck and Sikkink 1998). Transnational NGOs not only empower domestic actors by providing information, expertise, moral support, and financial assistance. They also enable societal actors to mobilize support at the international level to increase external pressure on their repressive governments (cf. Risse, Ropp, and Sikkink 1999). Like the European Commission, international organizations are an authoritative venue for non-state actors to challenge state violations of International Law. They provide new political opportunities to societal actors encouraging their connections with others like themselves and offering resources that can be used in intra-national and transnational conflict (Rogowski 1989; Haas 1989; Breitmeier and Wolf 1995).

While policy misfit and societal mobilization prove to be important factors in explaining differing degrees of compliance with law beyond the nation-state, be it European or international, the Pull-and-Push Model does not constitute a proper compliance theory. The model identifies just one causal mechanism of compliance. It defines conditions under which compliance problems are likely to occur and specifies the degree of compliance we may expect. The argument is grounded in a rationalist 'logic of consequentialism' (March and Olsen 1998), which treats actors as rational, goal-oriented and purposeful. Actors engage in strategic interactions using their resources to maximize their utilities on the basis of given, fixed and ordered preferences. They follow an instrumental rationality by weighing the costs and benefits of different strategy options taking into account the (anticipated) behavior of other actors. From this perspective, compliance problems emerge because 'misfitting' polices impose adaptational costs that actors seek to avoid. Their resistance can be overcome by reducing compliance costs through positive incentives (for instance, financial transfers) and by increasing the costs of non-compliance through (the threat of) sanctions (for instance, financial penalties, loss of electoral support). Through mobilizing pressure from below and above, domestic actors change the cost-benefit calculations of policy-makers and public authorities in favour of compliance, essentially by increasing the costs of non-compliance.

There are other ways of conceptualizing policy misfit and mechanisms that lead to compliance despite the high costs involved (cf. Börzel 2002a; Börzel and Risse 2002). The sociological 'logic of appropriateness' (March and Olsen 1998) argues that actors are guided by collective understandings of what constitutes proper, that is, socially accepted behavior in a given rule structure rather than by strategic cost-benefit calculations. Collective understandings and intersubjective meanings influence the ways in which actors define their goals and what they perceive as 'rational' action. Rather than maximizing their subjective desires, actors seek to fulfill social expectations. Compliance involves the internalization of new norms and rules through processes of socialization and habitualization, as a result of which actors may change their interests and identities. From this perspective, compliance problems also emerge because norms and rules are 'inconvenient'. But the emphasis is not so much put on the compliance costs caused by 'misfitting' policies but on their poor resonance with cognitive and ideational structures at the domestic level. The more European or international norms resonate with collectively shared beliefs and identities, the more likely they are internalized and complied with, and vice versa. Resistance against 'misfitting' norms can be overcome by processes of social learning and persuasion. Domestic actors do not so much pressure public actors into compliance by changing their cost-benefit calculations. Rather they act as norm entrepreneurs engaging opponents of compliance in a (public) discourse on why compliance with a ('misfitting') norm constitutes appropriate behaviour. For states, for instance, which perceive themselves as modern and liberal, violations of human rights norms can be framed as seriously contradicting their identity. Likewise, non-compliance with environmental regulations is little appropriate for states that pride themselves as environmental leaders. Using moral arguments and strategic constructions, domestic actors seek to persuade actors to redefine their interests and identities in processes of social learning (Finnemore 1996; Checkel 2001a; Risse, Ropp, and Sikkink 1999).[1]

Rationalist and sociological approaches identify different mechanisms of compliance but they are by no means mutually exclusive. Although the empirical study in this book supports the rationalist logic, it provides some ground for a more sociological reasoning. For instance, the German resistance against the Access to Information Directive is motivated by cognitive rather than material or political costs. The Directive does not resonate with the German state tradition, where transparency and broad access to public files are considered inappropriate. The study presents little evidence for the existence of norm entrepreneurs that seek to persuade Spanish and German policy-makers and public authorities into compliance with 'misfitting' EU policies. Yet, the very absence of epistemic communities and advocacy coalitions may help to explain why compliance is often such a tedious process. If compliance with 'misfitting' policies largely depends on pressure from below and above, sustained compliance is difficult to ensure. Since actors' interests remain unchanged, they are likely to fall back into non-compliant behaviour as soon as pressure is relaxed. This is particularly problematic for practical application where pressure is directed against isolated cases and institutional changes are less effective in affecting actors' behaviour. In other words, the pull-and-push mechanism may help to bring about formal changes in legal and administrative institutions. It is less likely to have an enduring effect on actors' behaviour since it leaves actors' interests and identities unchanged. Finally, the effect of public shaming, which has been identified as an important pull-and-push factor, is difficult to explain within a purely rationalist perspective. If adaptational costs are

---

[1]   The literature has identified two main types of norm- and idea-promoting agents. Epistemic communities are networks of actors with an authoritative claim to knowledge and a normative agenda (Haas 1992). They legitimate new norms and ideas by providing scientific knowledge about cause-and-effect relationships. Advocacy or principled issue networks are bound together by shared beliefs and values rather than by consensual knowledge (Keck and Sikkink 1998). They appeal to collectively shared norms and identities in order to persuade other actors to reconsider their goals and preferences.

high, public actors should not care that much about damages to their reputation, which ultimately rests on concerns about appropriate behaviour.

In sum, the Pull-and-Push Model presents one compliance mechanism, which draws on a rationalist understanding of actors' behaviour. It does not deny the relevance of alternative causal pathways to compliance. On the theoretical level, the sociological 'logic of appropriateness' may contradict the rationalist 'logic of appropriateness'. But the 'games real actors play' (Scharpf 1997) usually combine the two theoretically distinct logics of social action. They often work simultaneously or dominate different stages of the compliance process. Instead of pitching one logic against the other, future research should focus on exploring how the two pathways and causal mechanisms relate to each other (cf. Börzel 2002a; Börzel and Risse 2002). For instance, an 'inconvenient' norm that does not resonate with the domestic context of a state is likely to cause significant compliance costs. Likewise, actors that have to bear the costs of complying with a 'misfitting' policy are often inclined to present the policy as incompatible with domestic structures. Policy-makers that face pressure from above and below may be more open to processes of social learning and persuasion. Attempts of norm entrepreneurs to persuade public actors into compliance can trigger domestic mobilization. While the two logics of compliance can reinforce each other, one could also think of situations where they undermine each others' effects. Infringement proceedings have pushed Germany into formal compliance with 'misfitting' EU policies, but they have done little to increase the acceptance of these policies among German administrators, who feel that the Commission should concentrate its enforcement efforts on the laggard countries where environmental protection is generally poor.

A second challenge for future research lies in identifying conditions under which pressure from below and above is likely to emerge. Why have German and Spanish environmental groups largely refrained from mobilizing against the flawed implementation of the Drinking Water Directive? Why has the Commission remained inactive in the face of Spain's non-compliance with EU air pollution control and drinking water quality policies? Limited capacities on the one hand, and low problem pressure and issue salience, on the other hand, are certainly part of the explanation. Particularly Spanish environmental organizations and citizen groups are reluctant to invest their scarce resources in pulling down EU policies to the domestic level, which are of little concern to the public, and therefore, difficult to mobilize upon. In the absence of domestic mobilization, the Commission is less inclined to prosecute non-compliant member states, especially when it comes to practical application and enforcement. On the one hand, the Commission relies on domestic information to detect violations against European Law. On the other hand, the infringement proceedings require the Commission to formally respond to complaints lodged by citizens, companies, and public interest groups keeping them informed about the steps taken. Indeed, of the 12 cases, the Drinking Water Directive is the only one in which the Commission became active without systematic pulling-down attempts of domestic actors.

'New policy instruments' (NPI), such as the Environmental Impact Assessment and the Access to Information Directive, are designed to strengthen participatory

rights of citizens and public interest groups in the implementation of EU Environmental Law in order to improve compliance (Knill and Lenschow 2000b). But
domestic actors are not always capable of exploiting the new political opportunities
offered by European policies, since those very policies often give rise to
implementation problems in the first place (see below). Further research should
systematically explore the conditions under which domestic actors are both willing
and capable of pressuring policy-makers and public authorities towards compliance
with 'misfitting' policies. Particular attention may be paid to the role of
transnational actors in providing a link between mobilizing pressure from below
and above.

Making European Law Work: Increasing Pull and Push

The Pull-and-Push Model points to policy misfit and pressure from below and
above as the main explanatory factors in accounting for differing degrees of compliance with law beyond the nation state. Consequently, effective implementation
and compliance with European Law in both South and North may be enhanced by,
first, increasing the compatibility between European and domestic policies, second,
by strengthening the mobilization capacities of domestic actors, and, third, by
improving the monitoring and enforcement capacities of the Commission.

Given the diversities of national regulatory institutions, any European policy is
likely to cause substantial misfit for at least some member states. The issue is not
so much to avoid policy misfit but to make it more balanced. The 'leader-laggard'
dynamics in European environmental policy-making result in substantially more
misfit for the southern late-comers with their limited compliance capacities.
Having to implement costly European policies that do not always address their
most pressing environmental problems and tend to ignore their economic concerns
undermines both the capacity and the willingness of late-comers to comply. They
often perceive EU environmental policies as tailored to the economic interests and
environmental concerns of the 'rich north', which are imposed on the 'poor south'
(Aguilar Fernandez 1993: 232-233; Yearley, Baker, and Milton 1994). While the
regulatory contest among northern pace-setters has favoured progressive European
environmental policies, it undermines the effectiveness and the legitimacy of these
policies in those countries where effective environmental regulations are mostly
needed.

This is not to suggest that European environmental regulations should become
less ambitious. Yet, the EU has to come to terms with the 'dark side' of the leader-
laggard dynamics in European policy-making. If a high level of regulation is
deemed necessary and desirable at the European level, environmental late-comers
need to be given more time and flexibility to adapt (see also below). Moreover,
they have to be assisted in coping with the costs. Flexibility and capacity building
could be combined by making assistance conditional upon progress in implementation. In any case, financial instruments to support the implementation of EU
legislation would have to be stocked-up considerably. Finally, European policies

should also focus more on specific environmental problems of environmental late-comers helping these countries to develop and implement adequate solutions.

There are indications that the EU is becoming somewhat more sensitive to the concerns of the South. The Barcelona process within the EU Mediterranean Policy has started to address environmental problems in the region and may help the southern member states to develop their own policy solutions. Moreover, some EU policies address the problem of uneven compliance costs more explicitly. The Framework Convention on Climate Change entails a burden-sharing concept for implementing the emission cuts to which the EU committed itself under the Kyoto Protocol. Growing emissions of economically less advanced countries will be balanced by reductions of industrially more advanced countries. Likewise, the Integrated Pollution Prevention and Control Directive contains provisions that define the requirements for the 'best available technology' to reduce emissions in relation to 'the technical characteristics of the installation, its geographical location and local environmental conditions'. The flexibility clause allows to account for the diverging economic and environmental conditions in the member states. The need for such politics of flexibility, exemptions and derogations is likely to increase with the upcoming enlargement of the EU (see below).

Enhancing the pull-and-push effect is not an easy task either. Both pull and push crucially depend on the level of domestic mobilization against the deficient implementation of European environmental policies. Domestic mobilization tends to be less effective in southern countries, where environmental organizations and citizen groups have limited resources and environmental awareness is only emerging. As a result, domestic mobilization is often diffuse and, hence, less effective. While (trans)national environmental NGOs have become more influential, local groups are still weak. Resource constraints, however, can be a problem in northern member states, too. Spanish and German NGOs have been quite successful in mobilizing against the deficient transposition of 'misfitting' EU policies. Their concentrated lobbying activities at the national and European level have significantly improved formal compliance. Yet, formal compliance is increasingly less a problem. Nor does it necessarily lead to (more) effective practical application and enforcement as the cases of the Drinking Water Directive in Spain and Germany clearly show. Public authorities often manage to circumvent or water down European regulations in practical application and enforcement, as Germany does in case of Environmental Impact Assessment and Drinking Water Directives. Domestic actors are crucial for detecting issues of improper application since the willingness and the capacity of the Commission to control member state compliance beyond the level of complete and correct legal implementation is limited. But societal organizations and citizens rarely have the resources to act as effective watchdogs for the Commission. Moreover, the long and patient checking of daily practice is not as attractive to environmental groups as the denouncing of environmental scandals and the defense of more sensitive issues (Krämer 1997: 19-20). Compliance with costly policies, such as the Environmental Impact Assessment and the Access to Information Directives, will only slowly emerge, even though their legal implementation has improved. Progress will largely depend on the (continued) mobilization of domestic actors.

'New policy instruments' (NPI) aim to strengthen the mobilization capacities of domestic actors in EU environmental policy-making. But the findings of the book point at a serious limitation, if not a dilemma, of NPI in promoting member state compliance with European environmental policy through encouraging public participation in the implementation process. Their effect is likely to vary across the member states depending on the fit between NPI and existing administrative practice on the one hand, and on the relative strength (resources) of societal actors vis-à-vis public administration, on the other hand. In 'fit' countries, such as the UK or Denmark, NPI cause few implementation problems. But as public participation and access to information already constitute administrative practice, the added value of NPI for societal actors may be limited. In 'misfit' countries, such as Spain and Germany, NPI create their own implementation problems. Societal actors can only use NPI to pressure public administration if they have successfully pulled the policies down to the domestic level. Even for the relatively powerful German environmental organizations and citizen groups this is often not an easy task. In brief, NPI tend to strengthen the already strong (resourceful) societal actors while those who would really need to be empowered are not able to systematically use the NPI because they lack the necessary resources to invoke the new rights conferred to them. As the member states with the lowest compliance record are those with the weakest societal actors, the potential of NPI to improve member state compliance is indeed questionable. So far, NPI tend to cause compliance problems rather than help solving them. Overall, the possibilities of the European Union to strengthen domestic mobilization in southern societies appear to be rather limited. Supporting transnational networks that provide domestic actors with important resources including information, expertise, and funding could be one way of social empowerment, to which European actors can contribute.

Granting member states more flexibility in implementing European policies and strengthening their implementation capacities can help to relax the problem of policy misfit as a major cause of non-compliance. At the same time, exemptions, transition periods, and compensation payments may create additional incentives and opportunities for member states to avoid inconvenient European regulations. The five year derogation of the Large Combustion Plant Directive saved Spain some EUR 1.4 billion, but at the same time, the government did little to address the problem of air pollutant emission. Spanish authorities and companies have not always made appropriate use of the European funds they received for investing in environmental measures either.[2] If member states are allowed more leeway in adapting European policies to domestic arrangements, the capacity of the European Commission as the 'guardian of the treaties' to evaluate implementation outcomes has to be strengthened, particularly when it comes to practical application. Given

---

[2]   Interview with the Instituto para la Política Ambiental Europea, Madrid, 03/97.

the limited resources of domestic actors, the Commission cannot exclusively rely on mechanisms of decentralized monitoring and enforcement. European monitoring networks and European inspectors entitled to make regular on-site visits in the member states could be effective means to deter member states from abusing greater discretion in implementing European policies. But this would require the Commission to abandon its strategy of decentralized enforcement as the member states would have to accept a stronger presence of European authorities within their territories in return for more flexibility in implementation.

Facing the Challenge of Enlargement: Towards Flexibility

Eastern enlargement poses a major challenge to the effective implementation of European Environmental Law by reinforcing the dark side of the 'leader-laggard' dynamics in EU environmental policy-making. Like the Southern European member states, the Central and Eastern European candidate countries are environmental late-comers and policy takers. In the Agenda 2000, the European Commission stated upfront that 'none of the candidate countries can be expected to comply fully with the *acquis* in the near future, given their present environmental problems and the need for massive investments' (Commission of the European Communities 1997a: 67). The Pull-and-Push Model supports the pessimistic expectations of the Commission. First, the fit between the environmental legislation of the candidate countries with European Environmental Law is low. Like in the southern member states, the problem is not so much one of integrating 'misfitting' EU policies into a dense network of domestic regulations but to build up new regulatory structures. The candidate countries have to transpose 89 legal acts listed in the *Acquis Guide* by the time set for their accession (Commission of the European Communities 1997b). The costs of adopting the environmental *acquis* could easily amount to some three to five per cent of their GDP over the next years (OECD countries spend an average of between one and two per cent of their GDP on environmental protection, cf. Axelrod and Vig 1998).

Second, while compliance costs are high, implementation capacities of the candidate countries are severely limited. They face serious policy overload by having to transpose in a relatively short period of time a body of legislation that took more than 20 years to develop. Next to the legal implementation of the 89 pieces of EU legislation, the candidate countries have to establish a broader legal and administrative infrastructure to make European Environmental Laws work, including property rights, monitoring networks, and public participation requirements (OECD 1999: 67-69). Due to the high adaptational costs, the candidate states have asked for transition periods concerning about 25 per cent of the environmental *acquis* (Homeyer, Carius, and Bär 2000: 349-350). Such temporary exemptions may seriously constrain the ability of the Commission to push the candidate countries into compliance. Once they have joined, the Commission has no effective lever to exert pressure from above anymore. If the Central and Eastern Europeans follow the southern example, they will postpone implementation until the end of the transition period instead of using the time for gradual adaptation. This is particularly likely to happen if the EU does not help strengthening the compliance capacity of the candidate countries. So far, EU assistance programmes, such as PHARE or the Instrument for Structural Policies for Pre-Accession (ISPA), only cover a small fraction of the total compliance costs (OECD 1999: 128-129).

Third, environmental activism is limited in the candidate countries. Environmental mobilization played an important role in the 1989 overthrow of the socialist regimes in Central and Eastern Europe. But the 'environmental dividend' was lost after the first generation of post-socialist policy-makers had left office. Other actors were too weak to keep the momentum. Like in the Southern European countries, environmental protection ranks low on the political agenda where economic and social problems of transition dominate (Baker and Jehlicka 1998). Transnational NGOs are the most effective environmental organizations while local groups are still weak. Pull factors are further impaired by an enduring tradition of implementation failure (OECD 1999: 69) and low acceptance of state intervention (Carius, Homeyer, and Bär 1999).

Given the unfavourable constellation of pull-and-push factors, the prospects for compliance with European Environmental Law in Central and Eastern Europe are rather gloomy. If EU environmental policies are not sensitive to the economic and environmental concerns of late-comers, eastern enlargement may indeed split the EU into a 'leaders' and a 'laggards' camp with the latter clearly outweighing the former. So far, the 'leader-laggard' dynamics of EU environmental policy-making have not resulted in stable coalitions among the first-comers and the late-comers, respectively. The northern pace-setters rarely have been able to forge a stable alliance with each other. Rather, 'green' coalitions are formed on an issue-by-issue basis and remain liable to defection (Liefferink and Andersen 1998b: 262). The pace-setters may share similar 'norms of environmental behaviour' (Skjærseth 1994: 38). But their regulatory traditions are still rather incoherent. As a result, environmental leaders often pursue diverging policy preferences, which give rise to competition rather than cooperation among them.[3] Their regulatory contest has provided openings for tactical coalitions that cut across the 'leader-laggard' divide. Thus, the UK joined the southern member states in their opposition against the Large Combustion Plant Directive. Likewise, the southern member states supported the British position on the Integrated Pollution Prevention and Control Directive, which strongly opposed EU-wide emission standards but favoured the 'site-specific best practicable environmental option' (Héritier, Knill, and Mingers 1996: 244). Germany, the Netherlands, and the Nordic countries pushed for tighter exhaust standards for the Auto-Oil Directive, whereas France, together with Spain and Italy, rejected mandatory controls.

---

[3] Particularly the Nordic countries seem to try to avoid the impression that they form a 'green' block out of fear to become isolated in the EU (Liefferink and Andersen 1998b).

While northern first-comers have repeatedly teamed up with southern late-comers to oppose EU policy initiatives, southern foot-draggers might also support northern pace-setters as Greece did when it joined Denmark in pushing for more stringent clear air standards, although such coalitions have remained the exception. Finally, the southern late-comers have not been able to form stable alliances either. Like their northern counterparts, they view each other as temporary allies in fighting against costly regulations and in pressing for more European funds, respectively. At the same time, however, they compete among each other – not so much for up-loading domestic policy preferences to the EU level but for EU financial support in down-loading EU policies at the domestic level. Such competition has also precluded them from jointly pushing for EU policies that would address some of their commonly shared problems, such as deforestation.

The diversity of the member states and the dynamics of EU policy-making have so far prevented the emergence of a 'North-South conflict' in European environmental politics. Yet, eastern enlargement may significantly reinforce existing tensions changing the balance of power between leaders and laggards. The economies of the Central and Eastern European candidate countries are still weak. Like the Southern European member states, they seek to catch up with their wealthy Northern European counterparts. Moreover, the environmental situation in Central and Eastern Europe differs from Western Europe due to the different paths to socio-economic development after the Second World War. Central planning resulted in a marked contrast between heavily polluted 'hot spots' and relatively unspoiled areas with exceptionally high levels of biodiversity (Homeyer, Carius, and Bär 2000: 353-355). The industrially and environmentally advanced countries of the North will continue to push for the rapid adoption and progressive development of European environmental standards that address their most pressing environmental problems as well as prevent competitive disadvantages for their industries and create new export markets for their green industry, respectively. The economically weaker member states in the South and East, by contrast, may become increasingly reluctant to bear the costs of environmental policies, which they perceive as mostly directed to the economic interests and environmental problems of the North. In an enlarged Union, the first-comers will no longer be able to set the pace against the united opposition of the late-comers. Nor will the pace-setters be able to form a blocking minority against attempts of foot-draggers to lower existing environmental standards (Dehousse 1998: 147). Against this background, European policies are more likely to be shaped by the interests of the environmental late-comers rather than the first-comers.

Some have argued that candidate countries are less inclined to join the group of environmental laggards than is usually expected (Jehlicka 2001). Unlike the Southern Europeans, the Central and Eastern European countries used to pursue a more pro-active approach to environmental policy adopting regulations that went beyond existing EU legislation. Only since they entered the accession process have they developed into passive policy takers focusing on the adoption of the environmental *acquis*. Moreover, their common history and shared problems have not prompted the candidate countries to engage in any systematic coordination of their interests in the run-up process to accession. Nor have they sought to approach the Southern European late-comers for political support or the exchange of accession experience. Nevertheless, high compliance costs, limited implementation capacities, and low environmental mobilization suggest that the candidate countries will be more likely to side with the laggards than with the leaders.

If these expectations hold, increasing the flexibility of EU environmental regulations could be the only way to prevent a negative 'leader-laggard' dynamics driving the member states into a race to the bottom (Holzinger and Knoepfel 2000). The literature discusses three different ways of introducing flexibility in EU policy-making (cf. Homeyer, Carius, and Bär 2000: 363-366). Temporal flexibility involves the granting of transition periods and allows member states to adapt to European standards at different speeds (multiple speeds). Geographical flexibility sets different standards between groups of member states (variable geometry). Substantive flexibility, finally, provides for differentiation according to certain context factors, such ecological conditions or economic capacities.

More flexibility could carry the danger of an overall weakening of existing environmental standards through the rise of a multispeed 'greening of Europe' and a partial 'renationalization' of environmental policy-making (Homeyer, Carius, and Bär 2000). Nevertheless, it might be the only way to foster an albeit gradual improvement of environmental conditions in late-coming countries, both in the South and the East of Europe, particularly if transition periods and exemptions are coupled with European help for building up compliance capacities. Next to stocking-up accession funds, 'twinning' provides a more decentralized mechanism of capacity-building. As a part of the PHARE programme, it was established in the framework of the EU's 'intensified pre-accession strategy', created in 1997 (Commission of the European Communities 1997a). Member-state experts shall assist candidate countries in developing the legal and administrative structures required to effectively implement selected parts of the *acquis*. Civil servants who have specific knowledge in implementing certain EU policies are delegated to work inside the ministries and government agencies of the candidate countries, usually for one or two years. These 'pre-accession advisors' cooperate with a team of domestic experts on legal harmonization and the creation of proper implementation structures. While member states get reimbursement, the costs that arise in the preparation and implementation of twinning projects are often not fully covered.

Increasing the flexibility of European (environmental) legislation is not only in line with a trend that started back in the mid-1980s when the Single European Act was adopted (Homeyer, Carius, and Bär 2000: 364-366). It dove-tails with the political debate on a clearer delimitation of EU competencies called for by the Declaration on the Future of the Union annexed to the Niece-Treaty. Thinking about ways of how to grant the member states more leeway in implementing European policies seems to be more promising to ensure the effectiveness and legitimacy of European Law than drawing-up detailed lists of competencies meant to contain and roll-back European policy-making.

Bibliography

Aguilar Fernández, Susana. 1992. Políticas Ambientales y Diseños Institucionales en España y Alemania: La Comunidad Europea como Escenario de Negociación de una Nueva Area Política. Tesis Doctoral, Departemento de Sciencies Políticas, Universidad Complutense de Madrid, Madrid.

Aguilar Fernández, Susana. 1993. Corporatist and Statist Design in Environmental Policy: The Contrasting Roles of Germany and Spain in the European Community Scenario. *Environmental Politics* 2 (2): 223-247.

Aguilar Fernández, Susana. 1994. Spanish Pollution Control Policy and the Challenge of the European Union. *Regional Politics and Policy* 4 (1, special issue): 102-117.

Aguilar Fernández, Susana. 1997. *El reto del medio ambiente. Conflictos e intereses en la politica medioambiental europea.* Madrid: Alianza Universidad.

Alonso García, Enrique. 1994. *El Derecho Ambiental de la Comunidad Europea.* 2 vols. Vol. 2. Madrid: Civitas.

Andersen, Mikael Skou, and Duncan Liefferink. 1997. Introduction: The Impact of the Pioneers on EU Environmental Policy. In *European Environmental Policy. The Pioneers*, edited by M. S. Andersen and D. Liefferink. Manchester: Manchester University Press, 1-39.

Audretsch, H., ed. 1986. *Supervision in European Community Law.* New York: Elsevier.

Axelrod, Regina S., and Norman J. Vig. 1998. The European Union as an Environmental Governance System. In *The Global Environment. Institutions, Law, and Policy*, edited by N. J. Vig and R. S. Axelrod. Washington D.C.: CQ Press, 72-97.

Baker, Susan, and Petr Jehlicka, eds. 1998. *Dilemmas of Transition: The Environment, Democracy and Economic Reform in East Central Europe.* Illford: Frank Cass.

Baldock, D., and Long, T. 1987. *The Mediterranean Environment Under Pressure: The Influence of the CAP on Spain and Portugal and the "IMPs" D in France, Greece and Italy.* London: Institute for European Environmental Policy.

Banfield, Edward C. 1958. *The Moral Basis of a Backward Society.* Glencoe, ILL: Free Press.

Baumgartner, Frank, and Barry Jones. 1991. Agenda dynamics and instability in American politics. *Journal of Politics* 53 (4): 1044-1073.

Becker, Stefan. 1995. Das Schweigen der Ämter. *Öko-Test* (8): 31-39.

Bennett, Graham, ed. 1991. *Air Pollution Control in the European Community Implementation of the EC Directives in the Twelve Member States.* London: Graham and Trotman.

Benz, Arthur. 1984. *Kooperative Verwaltung. Funktionen, Voraussetzungen und Folgen*. Baden-Baden: Nomos.

Böckenförde, Ernst-Wolfgang. 1980. Sozialer Bundesstaat und parlamentarische Demokratie. In *Politik als globale Verfassung. Aktuelle Probleme des nationalen Verfassungsstaates. Festschrift für Friedrich Schäfer*, edited by J. Jekewitz, M. Melzer, and W. Zeh. Opladen: Westdeutscher Verlag, 182-199.

Boehmer-Christiansen, Sonja, and Jim Skea. 1991. *Acid Politics: Environmental and Energy Policies in Britain and Germany*. London: Belhaven Press.

Böhm-Amtmann, Edeltraud. 1997. 'Umweltpakt Bayern. Miteinander die Umwelt schützen'. EG-Öko-Audit-Verordnung und Substitution von Ordnungsrecht. *Zeitschrift für Umweltrecht* (4): 178-188.

Bohne, Eberhard. 1981. *Der informelle Rechtsstaat*. Berlin: Duncker & Humblot.

Börzel, Tanja A. 1999. The Domestic Impact of Europe. Institutional Adaptation in Germany and Spain. PhD Thesis, Department of Social and Political Science, European University Institute, Florence.

Börzel, Tanja A. 2001. Non-Compliance in the European Union. Pathology or Statistical Artefact? *Journal of European Public Policy* 8 (5): 803-824.

Börzel, Tanja A. 2002a. Non-State Actors and the Provision of Common Goods. Compliance with International Institutions. In *Common Goods: Reinventing European and International Governance*, edited by A. Héritier. Lanham, MD: Rowman & Littlefield, 155-178.

Börzel, Tanja A. 2002b. *States and Regions in the European Union. Institutional Adaptation in Germany and Spain*. Cambridge: Cambridge University Press.

Börzel, Tanja A., and Thomas Risse. 2002. Die Wirkung Internationaler Institutionen: Von der Normanerkennung zur Normeinhaltung. In *Regieren in internationalen Institutionen*, edited by M. Jachtenfuchs and M. Knodt. Opladen: Leske + Budrich, 141-182.

Breitmeier, Helmut, and Klaus Dieter Wolf. 1995. Analysing Regime Consequences: Conceptual Outlines and Environmental Explanations. In *Regime Theory and International Relations*, edited by V. Rittberger. Oxford: Clarendon Press, 339-360.

Brysk, Alison. 1993. From Above and From Below: Social Movements, the International System, and Human Rights in Argentina. *Comparative Political Studies* 26 (3): 259-285.

Bugdahn, Sonja. 2001. Developing Capacity against Tradition: The Implementation of the EU Environmental Information Directive in Germany, Great Britain and Ireland. PhD Thesis, Department of Political and Social Sciences, European University Institute, Florence.

Bulmer, Simon, and William Paterson. 1987. *The Federal Republic of Germany and the European Community*. London: Allen & Unwin.

Bundesministerium für Umwelt, Naturschutz und Reaktorsicherheit, ed. 1997. *Umweltgesetzbuch (UGB-KomE). Entwurf der Unabhängigen Sachverständigenkommission zum Umweltgesetzbuch beim Bundesministerium für Umwelt, Naturschutz und Reaktorsicherheit*. Berlin: Duncker & Humblot.

Burstein, Paul. 1999. Social Movements and Public Policy. In *How Social Movements Matter*, edited by M. Giugni, D. McAdam and C. Tilly. Minneapolis, London: University of Minnesota Press, 3-21.

Caporaso, James A., and Joseph Jupille. 2001. The Europeanization of Gender Equality Policy and Domestic Structural Change. In *Transforming Europe. Europeanization and Domestic Change*, edited by M. Green Cowles, J. A. Caporaso and T. Risse. Ithaca. NY: Cornell University Press, 21-43.

Carius, Alexander, Ingmar von Homeyer, and Stefani Bär. 1999. The Eastern Enlargement of the European Union and Environmental Policy: Challenges, Expectations, Speeds and Flexibility. In *Environmental Policy in a European Union of Variable Geometry? The Challenge of the Next Enlargement*, edited by K. Holzinger and P. Knoepfel. Basel: Helbig & Lichtenhahn, 141-180.

Checkel, Jeffrey. 1997. International Norms and Domestic Politics: Bridging the Rationalist-Constructivist Divide. *European Journal of International Relations* 3 (4): 473-495.

Checkel, Jeffery T. 2001a. Why Comply? Social Learning and European Identity Change. *International Organization* 55 (3): 553-588.

Checkel, Jeffrey T. 2001b. The Europeanization of Citizenship? In *Transforming Europe. Europeanization and Domestic Change*, edited by M. Green Cowles, J. A. Caporaso and T. Risse. Ithaca, NY: Cornell University Press, 180-197.

Choy i Tarrés, Antoni. 1990. Acción pública de la Generalidad de Cataluña en materia de medio ambiente. *Autonomies* (12): 189-209.

Choy i Tarrés, Antoni. 1992. Competencias y funciones del municipio en materia de medio ambiente. *Autonomies* (15): 77-98.

Cichowski, Rachel A. 1998. Integrating the Environment: The European Court and the Construction of Supranational Policy. *Journal of European Public Policy* 5 (3): 387-405.

Collins, Ken, and David Earnshaw. 1992. The Implementation and Enforcement of European Community Legislation. *Environmental Politics* 1 (4): 213-249.

Commission of the European Communities. 1984. *First Annual Report to the European Parliament on Commission Monitoring of the Application of Community Law (1983), COM (84) 181 final*. Brussels: Commission of the European Communities.

Commission of the European Communities. 1990. *Seventh Annual Report to the European Parliament on Commission Monitoring of the Application of Community Law (1989), COM (90) 288 (final)*. Brussels: Commission of the European Communities.

Commission of the European Communities. 1991. Monitoring of the Application by Member States of Environment Directives. Annex C to the Eighth Annual Report to the European Parliament on the Application of Community Law. *Official Journal of the European Communities* C 338 (31.12.1991).

Commission of the European Communities. 1993a. *Report from the Commission of the implementation of Directive 85/337/EEC on the assessment of the effects of certain public and private projects on the environment*. Brussels: Commission of the European Communities.

Commission of the European Communities. 1993b. *Tenth Annual Report on the Monitoring of the Application of Community Law (1992), COM (93) 320 final.* Brussels: Commission of the European Communities.

Commission of the European Communities. 1996. *Thirteenth Annual Report on Monitoring the Application of Community Law (1995), COM (96) 600 final.* Brussels: Commission of the European Communities.

Commission of the European Communities. 1997a. *Commission Staff Working Paper: Guide to the Approximation of European Union Environmental Legislation, SEC (97) 1608.* Brussels: Commission of the European Communities.

Commission of the European Communities. 1997b. *Agenda 2000, COM (97) 2000.* Luxembourg: Office of Official Publications of the Commission of the European Communities.

Commission of the European Communities. 2000. *Seventeenth Annual Report on Monitoring the Application of Community Law (1999), COM (2000) 92 final.* Brussels: Commission of the European Communities.

Cortell, Andrew P., and James W. Jr. Davis. 1996. How do International Institutions Matter? The Domestic Impact of International Rules and Norms. *International Studies Quarterly* 40: 451-478.

Craig, P.P. 1993. Francovitch. Remedies and the Scope for Damages Liability. *Law Quarterly Review* 109: 595-621.

Craig, P.P. 1997. Once More Unto the Breach: The Community, the State and Damages Liability. *Law Quarterly Review* 113: 67-94.

Cremer, Wolfram, and Andreas Fishan. 1998. New Environmental Instruments in Germany. In *New Instruments for Environmental Policy in the EU*, edited by J. Golub. London: Routledge, 55-85.

De la Torre, Maria, and Cliona Kimber. 1997. Access to Information on the Environment in Spain. *European Environmental Law Review* February: 53-62.

Dehousse, Renaud. 1998. Institutional Models for an Enlarged Union: Some Reflexions on a Non-Debate. In *An Even Larger Union? The Eastern Enlargement in Perspective*, edited by R. Dehousse. Baden-Baden: Nomos, 143-155.

Duina, Francesco G. 1997. Explaining Legal Implementation in the European Union. *International Journal of the Sociology of Law* 25 (2): 155-179.

Dyson, Kenneth. 1980. *The State Tradition in Western Europe: A Study of an Idea and Institution.* Oxford: Martin Robertson.

Eder, Klaus, and Maria Kousis, eds. 2001. *Environmental Politics in Southern Europe.* Dordrecht/Boston/London: Kluwer Academic Publishers.

Egeberg, Morten. 2001. How Federal? The Organizational Dimension of Integration in the EU (and Elsewhere). *Journal of European Public Policy* 8 (5): 728-746.

Engel, Rüdiger. 1992. Der freie Zugang zu Umweltinformationen nach der Informationsrichtlinie der EG und der Schutz von Rechten Dritter. *Neue Zeitschrift für Verwaltungsrecht* (2): 111-114.

Erbguth, Wilfried. 1991. Das UVP-Gesetz des Bundes: Regelungsgehalt und Rechtsfragen. *Die Verwaltung* (3): 283-323.

Erbguth, Wilfried. 1999. *Zur Vereinbarkeit der jüngeren Deregulierungsgesetzgebung im Umweltrecht mit dem Verfassungs- und Europarecht - am Beispiel des Planfeststellungsrechts.* Baden-Baden: Nomos.

Erichsen, Hans-Uwe. 1992. Der freie Zugang zu Informationen über die Umwelt. *Neue Zeitschrift für Verwaltungsrecht* (2): 409-419.

Escobar Gómez, Gabriel. 1994. Evaluación de Impacto Ambiental en España: resultados prácticos. *Ciudad y Territorio* 2 (102): 585-595.

Evans, A. C. 1979. The Enforcement Procedure of Article 169 EEC: Commission Discretion. *European Law Review* 4 (6): 442-456.

Finnemore, Martha. 1996. *National Interests in International Society.* Ithaca: Cornell University Press.

Finnemore, Martha, and Kathryn Sikkink. 1998. International Norm Dynamics and Political Change. *International Organization* 52 (4): 887-917.

Font, Nuria. 1996. La Europeización de la política ambiental en España. Un estudio de implementación de la directiva de evaluación de impacto ambiental. PhD thesis, Departamento de Ciencia Política y Derecho Público, Universität Autonóma de Barcelona, Barcelona.

Font, Nuria, and Francesc Morata. 1998. Spain: Environmental Policy and Public Administration. A Marriage of Convenience Officiated by the EU? In *Governance and Environment in Western Europe. Politics, Policy and Administration*, edited by K. Hanf and A.-I. Jansen. Essex: Longman, 208-229.

From, Johan, and Per Stava. 1993. Implementation of Community Law: The Last Stronghold of National Control. In *Making Policy in Europe. The Europeification of National Policy-Making*, edited by S. S. Andersen and K. A. Eliassen. London et al.: Sage, 55-67.

Gebers, Betty. 1993. Activities concerning Access to Information. *Newsletter of the Environmental Law Network International (ELNI)* 1/1993: 3.

Gebers, Betty. 1996. Germany. In *Access to Environmental Information in Europe. The Implementation and Implications of Directive 90/313/EEC*, edited by R. Hallo. London, The Hague, Boston: Kluwer Law, 95-110.

Gran, Karl-Heinz. 1989. Die Grenzwerte der Trinkwasserverordnung. *Bau Intern* (7): 116-118.

Grau i Creus, Mireia. 2001. When National Policy-Making Met the Comunidades Autonomas. Strategies and Political Pressure of the CA Governments towards National Institutions. PhD thesis, Department of Social and Political Science, European University Institute, Florence.

Grote, Jürgen R. 1997. *Interorganizational Networks and Social Capital Formation in the South of the South.* EUI Working Paper, RSC 97/38. Florence: European University Institute, Robert Schuman Centre for Advanced Studies.

Haas, Peter M. 1989. Do Regimes Matter? Epistemic Communities and Mediterranean Pollution Control. *International Organization* 43 (3): 377-403.

Haas, Peter M. 1992. Introduction: Epistemic Communities and International Policy Coordination. *International Organization* 46 (1): 1-36.

Haas, Peter M. 1993. Protecting the Baltic and the North Seas. In *Institutions for the Earth: Sources of Effective International Environmental Protection*, edited by P. M. Haas, R. O. Keohane and M. A. Levy. Cambridge, MASS: MIT Press, 133-182.

Haas, Peter M. 1998. Compliance with EU Directives: Insights from International Relations and Comparative Politics. *Journal of European Public Policy* 5 (1): 17-37.

Haigh, Nigel, ed. 2001. *Manual of Environmental Policy. The EU and Britain.* London: Institute for European Environmental Policy and Elsevier Science.

Hanf, Kenneth, and Egbert van de Gronden. 1998. The Netherlands: Joint Regulation and Sustainable Development. In *Governance and Environment in Western Europe. Politics, Policy and Administration*, edited by K. Hanf and A.-I. Jansen. Essex: Longman, 152-180.

Hanf, Kenneth, and Alf-Inge Jansen, eds. 1998. *Governance and Environment in Western Europe. Politics, Policy and Administration.* Essex: Longman.

Hartkopf, Günter, and Eberhard Bohne. 1983. *Umweltpolitik I: Grundlagen, Analysen und Perspektiven.* Opladen: Westdeutscher Verlag.

Haverland, Markus. 1999. *National Autonomy, European Integration and the Politics of Packaging Waste.* Amsterdam: Thela Thesis.

Heinelt, Hubert, Tanja Malek, Nicola Staeck, and Annette E. Töller. 2001. Environmental Policy: The European Union and a Paradigm Shift. In *European Union Environment Policy and New Forms of Governance. A Study on the Implementation of the Environmental Impact Assessment Directive and the Eco-Management and Audit Scheme Regulation in three Member States*, edited by H. Heinelt, T. Malek, R. Smith and A. E. Töller. Aldershot: Ashgate, 1-32.

Héritier, Adrienne. 1994. 'Leaders' and 'Laggards' in European Clean Air Policy. In *Convergence or Diversity? Internationalization and Economic Policy Response*, edited by B. Unger and F. v. Waarden. Aldershot: Avebury, 278-305.

Héritier, Adrienne. 1996. The Accommodation of Diversity in European Policymaking and its Outcomes: Regulatory Policy as a Patchwork. *Journal of European Public Policy* 3 (2): 149-176.

Héritier, Adrienne, Christoph Knill, and Susanne Mingers. 1996. *Ringing the Changes in Europe. Regulatory Competition and the Redefinition of the State: Britain, France, Germany.* Berlin, New York: De Gruyter.

Héritier, Adrienne, Christoph Knill, Susanne Mingers, and Martina Becka. 1994. *Die Veränderung von Staatlichkeit in Europa. Ein regulativer Wettbewerb. Deutschland, Großbritannien, Frankreich.* Opladen: Leske + Budrich.

Hey, Christian, and Uwe Brändle. 1992. *Umweltverbände und EG. Handlungsmöglichkeiten der Umweltverbände für die Verbesserung des Umweltbewußtseins und der Umweltpolitik in der Europäischen Gemeinschaft.* Opladen: Westdeutscher Verlag.

Hirsch, Frank. 1977. *The Social Limits to Growth.* London: Routledge.

Holzinger, Katharina, and Peter Knoepfel. 2000. The Need for Flexibility: European Environmental Policy on the Brink of Eastern Enlargement. In *Environmental Policy in a European Union of Variable Geometry. The Challenge of the Next Enlargement*, edited by K. Holzinger and P. Knoepfel. Basel: Helbing & Lichtenhahn, 3-35.

Homeyer, Ingmar von, Alexander Carius, and Stefani Bär. 2000. Flexibility or Renationalization: Effects of Enlargement on EC Environmental Policy. In *Risks, Reform, Resistance, and Revival*, edited by M. Green Cowles and M. Smith. Oxford: Oxford University Press, 347-368.

Hooghe, Marc. 1993. Too Little, Too Slow. *Planologisch Nieuws* 13 (2): 169-178.

Inglehart, Ronald. 1990. *Culture Shift in Advanced Industrial Societies*. Princeton, NJ: Princeton University Press.

Instituto para la Política Ambiental Europea, Madrid. 1997. *Manual de Política Ambiental Europea: la UE y España*. Madrid: Fundación MAPFRE.

Jänicke, Martin, and Helmut Weidner. 1997. Germany. In *National Environmental Policies. A Comparative Study of Capacity-Building*, edited by M. Jänicke and H. Weidner. Berlin et al.: Springer, 133-155.

Jehlicka, Petr. 2001. Environmental Implications of Eastern Enlargement of the EU: The End of Progressive Environmental Policy? *Manuscript*, Florence: European University Institute.

Jimenez, Manuel. 1997. *Consolidation through Institutionalization? The Dilemmas of the Spanish Environmental Movement in the 1990s*. Paper presented at the ECPR Joint Sessions of Workshops held in Bern, 27 February-4 March 1997.

Jönsson, Christer, and Jonas Tallberg. 1998. Compliance and Post-Agreement Bargaining. *European Journal of International Relations* 4 (4): 371-408.

Jordan, Andrew. 1999. The Implementation of EU Environmental Policy: A Policy Problem Without a Political Solution. *Environment and Planning C: Government and Policy* 17: 69-90.

Keck, Margaret, and Kathryn Sikkink. 1998. *Activists Beyond Borders: Advocacy Networks in International Politics*. Ithaca, NY: Cornell University Press.

Keohane, Robert O. 1984. *After Hegemony. Cooperation and Discord in the World Political Economy*. Princeton, NJ: Princeton University Press.

Keohane, Robert O., and Stanley Hoffmann. 1990. Conclusions: Community Politics and Institutional Change. In *The Dynamics of European Integration*, edited by W. Wallace. London: Pinter, 276-300.

Kimber, Cliona. 2000. Implementing European Environmental Policy and the Directive on Access to Environmental Information. In *Implementing EU Environmental Policy. New Directions and Old Problems*, edited by C. Knill and A. Lenschow. Manchester: Manchester University Press, 168-196.

Kitschelt, Herbert P. 1986. Political Opportunity Structures and Political Protest: Anti-Nuclear Movements in Four Democracies. *British Journal of Political Science* 16: 57-85.

Kloepfer, Michael. 1984. Gesetzeslähmung durch fehlende exekutive Vorschriften im Abwasserabgabengesetz? *Natur und Recht* (7): 258-263.

Knill, Christoph. 1997. The Impact of National Administrative Traditions on the Implementation of EU Environmental Policy. In *The Impact of National Administrative Traditions on the Implementation of EU Environmental Policy. Preliminary Research Report for the Commission of the European Union, DG XI*, edited by C. Knill. Florence: European University Institute, Robert Schuman Centre for Advanced Studies, 1-45.

Knill, Christoph. 1998. Implementing European Policies: The Impact of National Administrative Traditions. *Journal of Public Policy* 18 (1): 1-28.

Knill, Christoph. 2001. *The Transformation of National Administrations in Europe. Patterns of Change and Persistence.* Cambridge: Cambridge University Press.

Knill, Christoph, and Andrea Lenschow. 1998. The Impact of British and German Administrations on the Implementation of EU Environmental Policy. *Journal of European Public Policy* 5 (4): 595-614.

Knill, Christoph, and Andrea Lenschow. 2000a. Do New Brooms Really Sweep Cleaner? Implementation of New Instruments in EU Environmental Policy. In *Implementing EU Environmental Policy: New Directions and Old Problem*, edited by C. Knill and A. Lenschow. Manchester: Manchester University Press, 251-282.

Knill, Christoph, and Andrea Lenschow. 2000b. Introduction: New Approaches to Reach Effective Implementation - Political Rethoric or Sound Concepts? In *Implementing EU Environmental Policy: New Directions and Old Problem*, edited by C. Knill and A. Lenschow. Manchester: Manchester University Press, 3-8.

Knill, Christoph, and Andrea Lenschow. 2001. Adjusting to EU Environmental Policy: Change and Persistence of Domestic Administrations. In *Transforming Europe. Europeanization and Domestic Change*, edited by M. Green Cowles, J. A. Caporaso and T. Risse. Ithaca, NY: Cornell University Press, 116-136.

Kollman, Kelly. 2001. Convergence Through the Back Door? The Implementation and Use of Environmental Management Systems in Germany and the UK. Paper presented at the ECSA Seventh Biennial International Conference held in Madison, Wisconsin, May 31-June 2, 2001.

Kousis, Maria. 1994. Environment and the State in the EU Periphery: The Case of Greece. *Regional Politics and Policy* 4 (1, special issue): 118-135.

Kousis, Maria, Donatella della Porta, and Manuel Jimenez. 2001. Southern European Environmental Activism. Challenging the 'Laggards' Label. Paper presented at ECPR General Conference held in Canterbury, September 8-10.

Krämer, Ludwig. 1989. Enforcement of Community Legislation on the Environment. *Journal of the Regional Science Association* 137 (March): 243-248.

Krämer, Ludwig. 1997. *Focus on Environmental Law. Second Edition.* London: Sweet & Maxwell.

Kriesi, Hanspeter. 1991. *The Political Opportunity Structure of New Social Movements: Its Impact on Their Mobilization.* Berlin: WZB.

Kriesi, Hanspeter, Ruud Koopmans, Jan Willem Duyvendak, and Marco Giugni. 1995. *New Social Movements in Western Europe. A Comparative Analysis.* Minneapolis/London: University of Minnesota Press.

Krislov, S., Claus-Dieter Ehlermann, and Jospeh Weiler. 1986. The Political Organs and the Decision-Making Process in the United States and the European Community. In *Integration Through Law, Methods, Tools and Institutions: Political Organs, Integration Techniques and Judicial Process*, edited by M. Cappelletti, M. Seccombe and J. Weiler. Berlin: Gruyter, 3-112.

Kromarek, Pascale. 1987. *Vergleichende Untersuchung über die Umsetzung der EG-Richtlinien Abfall und Wasser*. Bericht des Instituts für Europäische Umweltpolitik im Auftrag des Umweltbundesamtes, 9/87. Berlin: Umweltbundesamt.

Kunzlik, Peter. 1996. Environmental Impact Assessment: Bund Naturschutz, Großkrotzenburg and The Commission's Retreat on the "Pipe-line" Point. *European Environmental Law Review* 5 (3): 88-93.

Küppers, Peter. 1994. Erlebnisse im Dschungel hessischer Umweltinformationen. Über den Umgang hessischer Behörden mit der EG-Umweltinformationsrichtlinie. *KGV-Rundbrief* (4): 22-26.

Küppers, Peter. 1995. Hand in Hand gegen Bürgerinnen und Bürger. Umweltministerium und Behörden verweigern in NRW berechtigte Informationswünsche. *KGV-Rundbrief* (2): 26.

Kurzer, Paulette. 2001. *Markets and Moral Regulation. Cultural Change in the European Union*. Cambridge: Cambridge University Press.

La Spina, Antonio, and Giuseppe Sciortino. 1993. Common Agenda, Southern Rules: European Integration and Environmental Change in the Mediterranean States. In *European Integration and Environmental Policy*, edited by J. D. Liefferink, P. D. Lowe and A. P. J. Mol. London, New York: Belhaven, 217-236.

Lambrechts, Claude. 1996. Environmental Impact Assessment. In *European Environmental Law. A Comparative Perspective*, edited by G. Winter. Aldershot: Dartmouth, 63-79.

Länderarbeitsgemeinschaft Wasser, LAWA. 1997. *Bericht zur Grundwasserbeschaffenheit*. Berlin: Kulturbuchverlag.

Larrue, Corinne, and Lucien Chabason. 1998. France: Fragmented Policy and Consensual Implementation. In *Governance and Environment in Western Europe. Politics, Policy and Administration*, edited by K. Hanf and A.-I. Jansen. Essex: Longman, 60-81.

Lenius, Thomas. 1994. BUND Untersuchung zum Umweltinformationsrecht in Deutschland am Beispiel der Störfall-Verordnung. *Unpublished study*.

Lenius, Thomas, and Felix Ekhardt. 1995. Umweltinformationsgesetz und Grundwassersanierung. *Unpublished study*.

Lenschow, Andrea. 1997. The Implementation of EU Environmental Policy in Germany. In *The Impact of National Administrative Traditions on the Implementation of EU Environmental Policy. Preliminary Research Report for the Commission of the European Union, DG XI, April 1997*, edited by C. Knill. Florence: European University Institute, Robert Schuman Centre for Advanced Studies.

Lewanski, Rudolf. 1993. Environmental Policy in Italy: From Regions to the EEC, a Multiple Tier Policy Game. Paper presented at ECPR Joint Sessions of Workshops held in Leiden, April 1993.

Lewanski, Rudolf. 1998. Italy: Environmental Policy in a Fragmented State. In *Governance and Environment in Western Europe. Politics, Policy and Administration*, edited by K. Hanf and A.-I. Jansen. Essex: Longman, 131-151.

Liefferink, Duncan. 1996. *Environment and the Nation State: The Netherlands, the EU and Acid Rain.* Manchester: Manchester University Press.

Liefferink, Duncan, and Mikael Skou Andersen. 1998a. Greening the EU: National Positions in the Run-up to the Amsterdam Treaty. *Enviornmental Politics* 7 (3): 66-93.

Liefferink, Duncan, and Mikael Skou Andersen. 1998b. Strategies of the 'Green' Member States in EU Environmental Policy-making. *Journal of European Public Policy* 5 (2): 254-270.

López Taracena, Antonio. 1995. *Evaluaciones de impacto ambiental y deslinde competencial.* Madrid: Ministerio de Obras Públicas, Transportes y Medio Ambiente.

Lübbe-Wolff, Gertrude. 1994. Die EG-Verordnung zum Umwelt-Audit. *Deutsches Verwaltungsblatt* 109 (7): 361-374.

Macrory, Richard. 1992. The Enforcement of Community Environmental Laws: Some Critical Issues. *Common Market Law Review* 29: 347-369.

Majone, Giandomenico. 1993. The European Community between Social Policy and Social Regulation. *Journal of Common Market Studies* 11 (1): 79-106.

Malek, Tanja, Hubert Heinelt, Jürgen Taeger, and Annette E. Töller. 2001. The Implementation of EMAS in Germany. In *European Union Environmental Policy and New Forms of Governance. A Study of the Implementation of the Environmental Impact Assessment Directive and the Eco-Management and Audit Scheme Regulation in three Member States,* edited by H. Heinelt, T. Malek, R. Smith and A. E. Töller. Aldershot: Ashgate, 107-118.

Malek, Tanja, and Annette E. Töller. 2001. The Eco-Management and Audit Scheme (EMAS) Regulation. In *European Union Environmental Policy and New Forms of Governance. A Study of the Implementation of the Environmental Impact Assessment Directive and the Eco-Management and Audit Scheme Regulation in three Member States,* edited by H. Heinelt, T. Malek, R. Smith and A. E. Töller. Aldershot: Ashgate, 43-55.

March, James G., and Johan P. Olsen. 1998. The Institutional Dynamics of International Political Orders. *International Organization* 52 (4): 943-969.

Martin Mateo, Ramon. 1977. *Derecho Ambiental.* Madrid: Instituto de Estudios de Administración Local.

Mendrinou, Maria. 1996. Non-Compliance and the European Commission's Role in Integration. *Journal of European Public Policy* 3 (1): 1-22.

Mergner, Richard. 1997. *UVP auf dem Prüfstand. Bilanz und Perspektiven aus Sicht des Bundes Naturschutz in Bayern e.V.* Nürnberg: Bund Naturschutz in Bayern.

Meyer-Rutz, Eckard. 1993. Die Umsetzung der EG-Richtlinie über den freien Zugang zu Informationen über die Umwelt in das deutsche Recht. In *Freier Zugang zu Umweltinformationen. Rechtsfragen im Schnittpunkt umweltpolitischer, administrativer und wirtschaftlicher Interessen,* edited by R. Breuer and e. al. Heidelberg: v. Decker, 5-11.

Müller, Edda. 1986. *Die Innenwelt der Umweltpolitik: sozial-liberale Umweltpolitik - (Ohn)macht durch Organisation?* Opladen: Westdeutscher Verlag.

Müller, Edda. 1994. Das Bundesumweltministerium – „Randbereich" der Bundes-regierung? Organisationsform mit dem Tasschenrechner. *Zeitschrift für Parlamentsfragen* (4): 611-619.

Müller-Brandeck-Bocquet, Gisela. 1996. *Die institutionelle Dimension der Umweltpolitik. Eine vergleichende Untersuchung zu Frankreich, Deutschland und der Europäischen Union.* Baden-Baden: Nomos.

Neyer, Jürgen, Dieter Wolf, and Michael Zürn. 1999. *Recht jenseits des Staates.* ZERP-Diskussionspapier, 1/99. Bremen: Zentrum für europäische Rechtspolitik.

Nolte, G. 1994. General Principles of German and European Administrative Law - A Comparison in Historical Perspective. *Modern Law Review* 57: 191-212.

OECD. 1994. *Environmental Performance Reviews: Portugal.* Paris: OECD Publications.

OECD. 1997. *Environmental Performance Reviews: Spain.* Paris: OECD Publications.

OECD. 1999. *Environment in the Transition to a Market Economy: Progress in Central and Eastern Europe and the New Independent States.* Paris: OECD Publications.

OECD. 2001. *Environmental Performance Reviews: Germany.* Paris: OECD Publications.

Ortega Alvarez, Luis. 1991. Organización del medio ambiente: La propuesta de una autoridad nacional para el medio ambiente. In *Estudios sobre la Constitución Espanola*, edited by S. Martin-Retortello and E. García Enterría. Madrid: Civitas, 3751-3800.

Partsch, Christoph. 1998. Brandenburgs Akteneinsichts- und Informationszugangsgesetz (AIG) - Vorbild für Deutschland? *Neue Juristische Wochenschrift* (35): 2559-2563.

Pehle, Heinrich. 1997. Germany: Domestic Obstacles to an International Forerunner. In *European Environmental Policy. The Pioneers*, edited by M. Skou Andersen and D. Liefferink. Manchester: Manchester University Press, 161-209.

Pehle, Heinrich, and Alf-Inge Jansen. 1998. Germany: The Engine in European Environmental Policy? In *Governance and Environment in Western Europe*, edited by K. Hanf and A.-I. Jansen. Harlow: Addison Wesley Longmann, 82-109.

Perdigó Solà, Joan. 1996. Organización administrativa del medio ambiente en los grandes municipios y areas de conurbanización. In *Derecho del medio ambiente y Administración local*, edited by J. Esteve Pardo. Madrid: Civitas, 371-394.

Poveda Gomez, Pedro. 1997. El régimen de distribución de las competencias ambientales entre las distintas administraciones públicas. Análisis legal y jurisprudencial. *Gaceta Jurídica de la Naturaleza y el Medio Ambiente* (141): 6-41.

Pridham, Geoffrey. 1994. National Environmental Policy-making in the European Framework: Spain, Greece and Italy in Comparison. *Regional Politics and Policy* 4 (1, Special Issue): 80-101.

Pridham, Geoffrey. 1996. Environmental Policies and Problems of European Legislation in Southern Europe. *South European Society and Politics* 1 (1): 47-73.

Pridham, Geoffrey, and Michelle Cini. 1994. Enforcing Environmental Standards in the European Union: Is There a Southern Problem? In *Environmental Standards in the EU in an Interdisciplinary Framework*, edited by M. Faure, J. Vervaele and A. Waele. Antwerp: Maklu, 251-277.

Putnam, Robert D. 1993. *Making Democracy Work: Civic Traditions in Modern Italy*. Princeton, NJ: Princeton University Press.

Rehbinder, Eckard, and Richard Stewart. 1985. *Environmental Protection Policy*. Vol. 2. Berlin: de Gruyter.

Reinhardt, Michael. 1992. Abschied von der Verwaltungsvorschrift im Wasserrecht? Zu den Auswirkungen der neuen Rechtsprechung des EuGH auf den wasserrechtlichen Vollzug in der Bundesrepublik Deutschland. *Die Öffentliche Verwaltung* (3): 102-110.

Risse, Thomas, and Stephen C. Ropp. 1999. Conclusions. In *The Power of Human Rights. International Norms and Domestic Change*, edited by T. Risse, S. C. Ropp and K. Sikkink. Cambridge: Cambridge University Press, 234-278.

Risse, Thomas, Stephen C. Ropp, and Kathryn Sikkink, eds. 1999. *The Power of Human Rights. International Norms and Domestic Change*. Cambridge: Cambridge University Press.

Röger, Richard. 1997. Das Recht des Antragstellers auf Wahl des Informationszugangs im Rahmen der Ermessensentscheidung nach § 4 Abs. 1 Satz 2 UIG. *Deutsches Verwaltungsblatt*: 885-888.

Rogowski, Ronald. 1989. *Commerce and Coalitions: How Trade Affects Domestic Political Alignments*. Princeton, NJ: Princeton University Press.

Rucht, Dieter. 1999. The Impact of Environmental Movements in Western Societies. In *How Social Movements Matter*, edited by M. Giugni, D. McAdam and C. Tilly. Minneapolis, London: University of Minnesota Press, 204-224.

Rüdiger, Wolfgang, and R. Andreas Krämer. 1994. Networks of Cooperation: Water Policy in Germany. *Environmental Politics* 3 (4): 52-79.

Sanchis Moreno, Fe. 1996. Spain. In *Access to Environmental Information in Europe. The Implementation and Implications of Directive 90/313/EEC*, edited by R. E. Hallo. London, The Hague, Boston: Kluwer Law, 225-248.

Sbragia, Alberta. 2000. Environmental Policy: The 'Push-Pull' of Policy-Making. In *Policy-Making in the European Union*, edited by H. Wallace and W. Wallace. Oxford: Oxford University Press, 293-316.

Scharpf, Fritz W. 1997. *Games Real Actors Play*. Boulder, CO: Westview.

Scherzberg, Arno. 1994. Freedom of Information – deutsch gewendet: Das neue Umweltinformationsgesetz. *Deutsches Verwaltungsblatt* (13): 733-745.

Schink, Alexander. 1994. Die Entwicklung des Umweltrechts im Jahre 1993 - Erster Teil. *Zeitschrift für angewandte Umweltforschung* 7 (2): 183-196.

Schink, Alexander. 1998. Die Umweltverträglichkeitsprüfung - eine Bilanz. *Natur und Recht* 4: 173-180.

Schneider, Jens-Peter. 1995. Öko-Audit als Scharniere in einer ganzheitlichen Regulierungsstrategie. *Die Verwaltung* 28 (3): 361-388.

Schneider, Volker. 2001. Institutional Reform in Telecommunications: The European Union in Transnational Policy Diffusion. In *Transforming Europe. Europeanization and Domestic Change*, edited by M. Green Cowles, J. A. Caporaso and T. Risse. Ithaca, NY: Cornell University Press, 60-78.

Schretzenmayr, Gernot. 1989. Wie gut ist unser Trinkwasser? Auf das Grundwasser kommt es in Bayern an. *Bau Intern* (7): 119-122.

Schwanenflügel, Matthias. 1993. Die Richtlinie über den freien Zugang zu Umweltinformationen als Chance für den Umweltschutz. *Die Öffentliche Verwaltung* (3): 95-102.

Sikkink, Kathryn. 1993. Human Rights, Principled Issue Networks, and Sovereignty in Latin America. *International Organization* 47 (3): 411-441.

Skjærseth, Jon Birger. 1994. The Climate Policy of the EC: Too Hot to Handle? *Journal of Common Market Studies* 32 (1): 25-45.

Smollich, Thomas. 1992. Umweltverwaltungen der Länder. In *Umwelt, Handwörterbuch*, edited by F.-J. Dreyhaupt, F.-J. Peine and G. Wittkämper. Berlin, New York: de Gruyter, 422-436.

Snyder, Francis. 1993. The Effectiveness of European Community Law. Institutions, Processes, Tools and Techniques. *Modern Law Review* (56): 19-54.

Spanou, Calliope. 1998. Greece: Administrative Symbols and Policy Realities. In *Governance and Environment in Western Europe. Politics, Policy and Administration*, edited by K. Hanf and A.-I. Jansen. Essex: Longman, 110-130.

Spindler, Edmund A. 1994. *UVP Beschwerde aus Brüssel*. Pressemitteilung, Hamm: UVP Förderverein.

SRU, Sachverständigen Rat für Umweltfragen. 1994. *Umweltgutachten 1994. Für eine dauerhaft-umweltgerechte Entwicklung*. Stuttgart: Metzler-Poeschel.

SRU, Sachverständigen Rat für Umweltfragen. 1996. *Unweltgutachten 1996*. Stuttgart: Metzler-Poeschel.

Staeck, Nicola, and Hubert Heinelt. 2001. The Implementation of EIA in Germany. In *European Union Environment Policy and New Forms of Governance. A Study on the Implementation of the Environmental Impact Assessment Directive and the Eco-Management and Audit Scheme Regulation in three Member States*, edited by H. Heinelt, T. Malek, R. Smith and A. E. Töller. Aldershot: Ashgate, 61-70.

Staeck, Nicola, Tanja Malek, and Hubert Heinelt. 2001. The Environmental Impact Assessment Directive. In *European Union Environment Policy and New Forms of Governance. A Study on the Implementation of the Environmental Impact Assessment Directive and the Eco-Management and Audit Scheme Regulation in three Member States*, edited by H. Heinelt, T. Malek, R. Smith and A. E. Töller. Aldershot: Ashgate, 33-42.

Tallberg, Jonas. 1999. *Making States Comply. The European Commission, the European Court of Justice and the Enforcement of the Internal Market*. Lund: Studentlitteratur.

Tarrow, Sidney. 1998. *Power in Movement. Social Movements and Contentious Politics*. Cambridge: Cambridge University Press.

172 *Environmental Leaders and Laggards in Europe*

Treisman, Daniel. 2000. The Cause of Corruption: A Cross-National Study. *Journal of Public Economics* (76): 399-457.

Turiaux, André. 1994. Das neue Umweltinformationsgesetz. *Neue Juristische Wochenschrift* (36): 2319-2324.

Ulbert, Cornelia. 1997. Ideen, Institutionen und Kultur. Die Konstruktion (inter-) nationaler Klimapolitik in der BRD und in den USA. *Zeitschrift für Internationale Beziehungen* 4 (1): 9-40.

Underdal, Arild. 1998. Explaining Compliance and Defection: Three Models. *European Journal of International Relations* 4 (1): 5-30.

Voigt, Rüdiger, ed. 1995. *Der kooperative Staat*. Baden-Baden: Nomos.

Waskow, Siegfried. 1997. *Betriebliches Umweltmanagement. Anforderungen nach der Audit Verordnung der EG und dem Umweltauditgesetz*. 2. Auflage ed. Heidelberg: C.F. Müller.

Wates, Jeremy. 1996. The UN ECE Guidelines and Draft Convention on Access to Environmental Information and Public Participation in Environmental Decisionmaking. In *Access to Environmental Information in Europe: The Implementation and Implications of Directive 30/313/EEC*, edited by R. Hallo. The Hague: Kluwer Law International, 121-174.

Weale, Albert. 1992. *The New Politics of Pollution*. Manchester: Manchester University Press.

Weale, Albert, Geoffrey Pridham, Michelle Cini, Dimitrios Konstadakopulos, Martin Porter, and Brendan Flynn. 2000. *Environmental Governance in Europe. An Ever Closer Ecological Union?* Oxford: Oxford University Press.

Weale, Albert, Geoffrey Pridham, Andrea Williams, and Martin Porter. 1996. Environmental Administration in Six European States: Secular Convergence or National Distinctiveness? *Public Administration* 74: 255-274.

Wegener, Bernhard. 1994. Wie es uns gefällt oder ist die unmittelbare Wirkung des Umweltrechts der EG praktisch? Die Rechtsprechung deutscher Gerichte zum Recht auf freien Zugang zu Umweltinformation. *Zeitschrift für Umweltrecht* (5): 232-236.

Weidner, Helmut. 1995. *25 Years of Modern Environmental Policy in Germany. Treading a Well-worn Path to the Top of the International Field.* WZB Discussion Paper, FS II: 95-301. Berlin: Wissenschaftszentrum Berlin.

Weiler, Joseph. 1988. The White Paper and the Application of Community Law. In *1992: One European Market?*, edited by R. Bieber, R. Dehousse, J. Pinder and J. H. H. Weiler. Baden-Baden: Nomos, 337-358.

Wey, Klaus-Georg. 1982. *Umweltpolitik in Deutschland. Kurze Geschichte des Umweltschutzes in Deutschland seit 1900*. Opladen: Westdeutscher Verlag.

Williams, R. 1994. The European Commission and the Enforcement of Environmental Law: An Invidious Position. *Yearbook of European Law* 14: 351-400.

Winter, Gerd. 1996. Freedom of Environmental Information. In *European Environmental Law. A Comparative Perspective*, edited by G. Winter. Aldershot: Dartmouth, 81-94.

Yearley, Steven, Susan Baker, and Kay Milton. 1994. Environmental Policy and Peripheral Regions of the European Union: An Introduction. *Regional Politics and Policy* 4 (1, Special Issue): 1-21.

Zito, Anthony R. 2000. *Creating Environmental Policy in the European Union.* London: Macmillan.

Index

Printed and bound by CPI Group (UK) Ltd, Croydon, CR0 4YY

21/10/2024

01777082-0014